Pruning & Training

Essential know-how and expert advice for gardening success

CONTENTS

UNDERSTANDING PRUNING AND TRAINING

Before you begin to prune and train your plants, it helps to understand why it is necessary. By following the correct techniques, you can be safe in the knowledge that you are pruning correctly and at the optimum time, and training growth in a way that will encourage it to thrive. Learning to select the correct tools to make clean cuts is also key to producing good results and keeping your plants healthy.

WHY PRUNE AND TRAIN?

Many trees, shrubs, and climbers will grow happily when left to their own devices, but pruning and training can be used in a huge variety of ways to influence their appearance and to enhance their ornamental qualities, allowing you to create and maintain a beautiful garden. Careful pruning and training increases the quality and quantity of flowers on many climbers and shrubs, which may then clothe walls and archways with their lush growth. Pruning is also vital to remove diseased or dead stems to keep plants healthy and can even trigger new vigour in tired overgrown specimens.

Naturally elegant Japanese maples (*Acer*) are pruned only minimally.

FOR PLANT HEALTH

Pruning and training can be used throughout a plant's life to promote good health and deal with any problems that may arise. Routine pruning to create an open shape, free from congested branches, not only makes a plant more attractive but also helps to prevent disease by improving airflow around the foliage and stopping branches rubbing together in the wind.

Damaged and dead growth provide an easy route for disease to enter a plant, and so are always removed first during pruning. Cutting back to healthy tissue using clean sharp tools halts the spread of any disease already present and allows plants to heal quickly, minimizing the chance of new infections and encouraging strong new growth.

TO SHAPE PLANTS

Most gardeners regularly trim a hedge to keep it to their desired shape, but other useful pruning and training methods can also be employed to manipulate the growth of garden plants, often to spectacular effect. To achieve the best results, begin pruning when plants are young, so that you can establish the balanced structure of branches at the heart of a specimen tree or shrub, or encourage certain shrubs to develop into "lollipop" standards. Many flowering and berry-bearing shrubs can also be trained to grow tightly against a wall, which produces a spectacular display and is an excellent way to accommodate large plants in small gardens.

Pruning plants in order to maintain their planned forms is always much easier and more rewarding than trying to tame the growth of vigorous climbers, shrubs, or trees that have outgrown their allotted space.

Train wisteria to maximize the impact of its cascading flowers.

Clipped topiary adds formal structure to a garden, whatever its size.

TO BOOST PERFORMANCE

Plants react to being cut back by producing fresh new growth, so make use of this response to get the best from your plants. Well-timed pruning and careful training can promote flowering, increasing both the quality and quantity of blooms on a single plant, sometimes followed by a brilliant display of decorative berries or hips on fruiting plants in autumn.

Regular pruning or trimming each spring can combat the tendency of some climbers and shrubs to become leggy or bare at the base as they age, encouraging compact bushy growth, which makes an attractive feature when the plants are not in flower.

Severe pruning is also frequently used to renovate a plant that has become overgrown. Removing its tangled old stems stimulates vigorous new growth and can revitalize a tired corner of the garden.

Encourage roses to bloom freely with the appropriate pruning and training.

Fluffy seedheads make an attractive feature once clematis flowers have faded.

Train climbing hydrangea (*Hydrangea petiolaris*) so it covers a shady wall.

FOR ATTRACTIVE YOUNG GROWTH

The young shoots of some trees and shrubs are valued for decorative qualities that are quite different from those of the mature plants. Specific pruning techniques, known as coppicing or pollarding, involve cutting older growth back hard – often every spring – to promote vigorous new stems (*see pp.28–29*). These shoots either carry large colourful leaves, such as the circular silvery juvenile foliage of gums (*Eucalyptus*), or boast vividly coloured bark, like the bold reds and oranges of some dogwood (*Cornus*) varieties, which lights up a border in winter. This severe pruning can also be a useful technique for limiting the size of large plants, allowing them to be grown successfully in small gardens.

TOP TIP NEVER MAKE A CUT WITHOUT A GOOD REASON. ALWAYS STEP BACK AND LOOK AT THE HEALTH, STRUCTURE, AND GROWTH OF THE WHOLE PLANT BEFORE EMBARKING ON ANY PRUNING.

NEED TO KNOW

• Pruning is cutting growth from a plant to improve its health, shape, or production of flowers and fruit.
• Training involves directing plant growth, usually by tying it into a supporting structure, to form an attractive branch framework and often to increase flowering.

Coppiced dogwoods (*Cornus*) produce a profusion of new "fiery" growth.

WHAT TO CONSIDER
BEFORE CUTTING

Pruning may seem daunting, but if you know your plants you can make your pruning cuts purposeful. Start by identifying each plant, then check the A–Z (*see pp.40–141*) for advice on when and how to prune it correctly. Consider what you want to achieve when you prune, how it could affect wildlife, and ensure that you have the right tools to carry out the job effectively and safely.

Promote flowering to attract beneficial insects, by pruning and training.

IS PRUNING THE SOLUTION?

Many trees and shrubs flourish when their beautiful natural forms are allowed to develop without any interference. However, sometimes a large shrub or tree has been given a cramped location or planted too close to its neighbours and will only look its best if given more space. Where such a plant then has to be pruned heavily every year to restrict its size, the better option may be to remove it. If you decide to replace it, carefully check the mature height and spread of any plant before you buy it.

Cornus kousa 'Snowboy' has a graceful natural habit and needs little pruning.

WHAT DO I WANT TO ACHIEVE?

Plants should be pruned only for a clear reason, so ask yourself what is your aim for each particular plant. The most important one is probably to keep your plant healthy so the prompt removal of dead, diseased, and damaged growth will always be your top priority. Then take a step back from each specimen to assess how pruning could influence and improve its shape. Formative pruning of young plants is critical to establish a clean trunk or attractive branch structure, while maintenance pruning keeps this shape balanced as plants mature.

Another motive for pruning plants is to promote flowering, fruiting, or other ornamental characteristics and, as these tasks are repeated annually, they will quickly become a familiar part of your gardening routine. You must always be careful to prune each plant at the correct time of year, to avoid removing developing flowers or fruits.

Shape young plants so that they mature into attractive specimens.

| TOP TIP REMOVE SHOOTS WITH LEAVES THAT ARE A DIFFERENT COLOUR TO THOSE ON THE REST OF THE PLANT IMMEDIATELY YOU SPOT THEM, BECAUSE VIGOROUS "REVERTED", PLAIN GREEN GROWTH ON VARIEGATED PLANTS WILL QUICKLY DOMINATE IF LEFT UNTOUCHED.

A good crop of berries adds welcome colour through autumn and winter, and will keep birds well fed.

Flower-laden shrubs are rich nectar sources for butterflies and other insects.

Nesting birds frequently make use of gardens; take care not to disturb them.

WILDLIFE

Garden hedges, trees, shrubs, and climbers are all invaluable resources for a huge variety of wildlife, providing food, shelter, and nesting sites for everything from tiny insects to birds and mammals. It is therefore important to consider the welfare of these creatures before you tackle any pruning, because many of them are beneficial in the garden and help control the numbers of common pests such as aphids and snails. In the UK, it is an offence to damage the nest of any bird while it is being built or in use, so any pruning of trees, shrubs, and hedges where breeding birds are active should be delayed until early autumn.

Routine pruning is often carried out specifically to enhance flowering performance and the profusion of nectar-rich blooms on plants such as lavender (*Lavandula*) and buddleja, which are well known magnets for butterflies and many other insects. You therefore need to ensure that you time pruning correctly so that you do not remove growth that will bear the next flowers and fruits, which will feed these insects and birds.

DO I HAVE THE RIGHT EQUIPMENT?

Clean cuts heal quickly and neatly, and these can only be made with tools appropriate for the size of stems or branches to be pruned (*see pp.16–17*). Secateurs will suffice for pruning stems up to pencil thickness, but to cut thicker or woody growth you require the extra leverage of loppers or a pruning saw. You also need to protect yourself from thorns and prickles by wearing thornproof gloves, and always put on safety glasses or goggles when trimming hedges or pruning climbers.

Never tackle a pruning job with inadequate or blunt tools because they will crush and tear the stems, which will then look tatty, take longer to heal, and be vulnerable to disease. You are also much more likely to be injured, so it is always best to wait to prune until you have access to suitable tools.

Use a sharp pruning saw to make accurate cuts through thick woody stems.

HOW TO MINIMIZE PRUNING

The amount of pruning required to keep your garden healthy and full of vibrant growth depends a great deal on the plants in your care. Ensure that any new plants are healthy and shapely in order to reduce the need for pruning while they are young. If you are looking for low-maintenance options, then there are many trees and shrubs that flourish with no regular pruning and even some that actively dislike any intervention. It is always worth including some of these plants in any garden, to help keep it manageable and as a reminder of how attractive the natural forms of plants can be.

Select specimens with uniform, evenly spaced shoots, like these rosemary plants.

START WITH A GOOD SPECIMEN

Choosing healthy, well-shaped young plants at the garden centre or nursery will help to ensure that they establish quickly and will probably save you having to prune to create an attractive form or to remove damaged shoots. Look for plants full of vigour, with unblemished leaves, or with plenty of plump buds on bare winter stems. Avoid leggy specimens, especially of shrubs that form dense mounds of foliage, because even with pruning this will be difficult to correct. Assess young trees to check that there is a balanced arrangement of branches and a clear trunk; buying specimens that have already been formatively pruned means that you will need to prune only later on, to maintain their structures. Where plants are displayed close together, always move them apart so you can carefully inspect each from all sides before purchasing.

CHOOSE NATURALLY NEAT PLANTS

Shrubs that develop tidy compact growth are perfect for small gardens and generally need only minimal pruning. This ability to thrive with little maintenance also makes them useful in areas where access for pruning is awkward, such as on a steep slope. Their dense mounds of small foliage are often evergreen, adding attractive structure to the garden, and many of these plants also produce masses of vibrant flowers or colourful berries. Look out for dwarf or compact varieties of larger shrubs, which are often ideal where space is at a premium and should remove any subsequent need to prune to restrict their size.

COMPACT SHRUBS Gaultheria • Hebe • Mexican orange blossom (*Choisya*) • *Potentilla fruticosa* varieties • Rock rose (*Cistus*) • Skimmia • Sweet box (*Sarcococca*)

Rock roses require no pruning to produce their mass of summer flowers.

Hebes spread gradually to form dense mounds of evergreen foliage.

SELECT SHAPELY SHRUBS AND TREES

Many trees and shrubs owe their popularity as garden plants to their distinctive and elegant natural forms. Interfering with their growth by carrying out anything more than essential pruning risks spoiling the shape for which they are valued, so they are best left to their own devices as much as possible. Ensure that these trees and shrubs are planted with enough space to spread their branches unhindered, perhaps as a specimen in a lawn or surrounded by lower-growing perennials.

To keep these plants healthy, always remove dead, diseased, and damaged wood along with any branch that grows in a wayward direction and upsets the plant's form.

PLANTS WITH NATURAL ELEGANCE *Cedrus deodara* 'Aurea' • Dogwood (*Cornus*) • Irish yew (*Taxus baccata* 'Fastigiata') • Japanese maples (*Acer japonicum, A. palmatum*) • Magnolia • Sumach (*Rhus*) • Witch hazel (*Hamamelis*)

Magnolias naturally develop a graceful airy arrangement of branches.

Eye-catching sumach is multi-stemmed without any need to shape it.

The distinctive shapes of some conifers are best left unpruned.

CONIFERS

The diverse habits and sizes of different conifers mean that you can find one to suit almost any situation in the garden. Most conifers require little pruning and training. In fact, the majority respond poorly to pruning, which makes it especially important to plant them in a location that suits their mature size. However, varieties of thuja, cypress (*Cupressus*), and yew (*Taxus*), along with Lawson cypress (*Chamaecyparis lawsoniana*), can all be trimmed annually with hedge trimmers or shears to check their spread and are ideal to plant as dense hedges.

PLANTS THAT DISLIKE BEING PRUNED

Some plants, notably daphnes and tree peonies (for example, *Paeonia delavayi, P. ludlowii,* and *P. rockii*) respond badly to pruning except clean cuts to remove damaged wood. A number of trees, including birches (*Betula*), Japanese maples (*Acer japonica, A. palmatum*), and magnolias, will bleed by oozing sap from any cuts made when they are coming into growth in late winter or spring. This problem can be avoided by carrying out any pruning on them only when growth is slowing or the plants are dormant, between late summer and midwinter.

Always check the pruning recommendations for each plant (*see the A–Z, pp.40–141*) before embarking on any renovation work because some, including most conifers, will fail to regrow if cut back into old wood.

Slow-growing daphnes often suffer from dieback after pruning.

Tree peonies are best left to create their own unique forms.

HOW PLANTS RESPOND TO TRAINING AND PRUNING

A single leading shoot has produced strong vertical growth on this conifer.

By understanding the ways in which plants grow and react to cuts and other types of manipulation, you can take the guesswork out of training and pruning. This will enable you to make your own assessment of the work that each plant requires and to carry it out with confidence. Stems can be cut or trained to control the vigour of new shoots, to promote dense bushy growth, or to encourage prolific flowering, to give you healthy plants and spectacular colour.

APICAL DOMINANCE

Young shoots on lateral (side) stems and main stems develop from behind a "terminal" bud at their tip and are the focus of a plant's growth. These buds are extremely influential, because they are able to exert what is called "apical dominance" thanks to hormones that flow down from the shoot tip and prevent the growth of the lateral buds along the stem.

Apical dominance often allows a single leading shoot to extend vertically, creating an upright form, which can be desirable in many trees. In other instances, however, pruning to shorten vigorous vertical shoots and break this apical dominance can improve the appearance and flowering performance of garden plants.

PRUNING TO REMOVE A DOMINANT SHOOT

By cutting a stem to remove its terminal bud you can stop the flow of hormones that suppress the growth of lateral buds lower down the stem, thus breaking apical dominance. This triggers the buds immediately below the pruning cut to produce strong lateral shoots along the stem. Pruning to break apical dominance therefore reduces upright growth in favour of branched stems, and so creates plants with a bushier habit, denser foliage, and a greater number of flowering shoots – all of which are desirable characteristics in many garden plants.

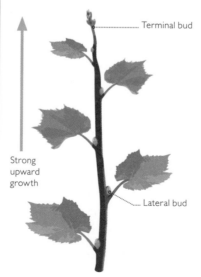

Terminal bud

Strong upward growth

Lateral bud

Before being cut, the dominant shoot extends growth vertically.

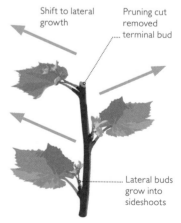

Shift to lateral growth

Pruning cut removed terminal bud

Lateral buds grow into sideshoots

After the dominant shoot is cut, the lateral buds form strong sideshoots.

HORIZONTAL TRAINING

Apical dominance can also easily be disrupted without making any pruning cuts. This alternative method involves training the naturally more vertical stems of climbers, shrubs, and trees so they are held horizontally, in order to reduce the quantity of hormones travelling along the stem from the apical bud and to stimulate the lateral buds to grow upwards. These shoots are ideal to clothe a wall or fence with an even covering of foliage. Horizontal training is also a great way to encourage flowering and fruiting, at the expense of vigorous leafy growth, and is often used on climbing roses and wisteria (see p.38).

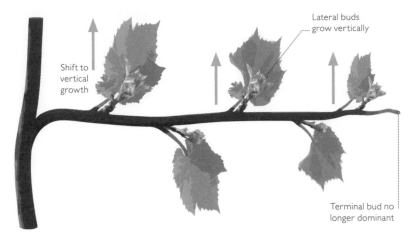

Shift to vertical growth

Lateral buds grow vertically

Terminal bud no longer dominant

Bending a stem horizontally and tying it in that position causes the lateral buds along its length to grow upwards evenly.

CONTROLLING PLANT VIGOUR

Pruning stimulates growth, but the vigour of these new shoots varies, depending on how hard the plant is pruned. Severe pruning promotes stronger growth than light pruning, because the more stems a plant loses the harder it will work to replace them. Therefore, you should prune a large or fast-growing plant only lightly to keep it in check, and you must fight any urge to cut it back hard to restrict its spread, because this will only result in further vigorous growth.

Conversely, when a plant is growing weakly the best thing to do is cut it back severely, to stimulate growth. This can seem counter-intuitive when you are new to pruning, but it quickly makes sense as you gain experience.

TOP TIP AFTER PRUNING, FEED THE PLANT WITH A BALANCED GRANULAR FERTILIZER AND MULCH THE SOIL AT ITS BASE WITH A THICK LAYER OF WELL-ROTTED COMPOST TO PROVIDE NUTRIENTS TO FUEL NEW GROWTH.

Cut back curry plants (Helichrysum) to promote compact flushes of new foliage.

Lightly prune vigorous climbing roses to avoid stimulating excessive growth.

Severely pruned trees often produce vertical growth, which needs removing.

PRUNING TOOLS

Good-quality pruning tools can be expensive, but they should last you a lifetime if kept clean, oiled, and sharp, and will repay your investment by making pruning a pleasure. Each tool is designed for a specific job, so buy or borrow all those that you need to enable you to maintain the plants in your garden without causing them damage or injuring yourself. Protective equipment is also essential to prevent thorns, whippy stems, and irritant sap coming into contact with your skin and eyes.

Hedging shears make quick work of many garden trimming tasks.

CUTTING TOOLS

Secateurs are the most versatile cutting tool and can be used to prune everything from soft green shoots to woody stems up to about 1cm (½in) in diameter. Choose a sturdy pair with a comfortable grip. Some secateurs have scissor-like "bypass" blades, which glide past one another to make clean precise cuts. Less common are "anvil" secateurs, where a sharp upper blade cuts down on to a flat lower blade. The latter are ideal for cutting through tough dead wood and thicker branches, but the crushing nature of the action means they are unsuitable for general pruning. Loppers have larger blades and long handles, which provide the extra leverage often required to cut through woody stems and small branches that are too thick for secateurs.

To remove more substantial branches on trees and mature shrubs, a pruning saw will be needed. It is much better for these tasks than a standard household saw, because its special thin cutting blade and handle are designed to fit easily between branches, while the curve of the blade prevents it becoming clogged.

Use topiary shears where you wish to create or fine-tune intricate designs.

TRIMMING TOOLS

Hedging shears are the ideal tool for trimming low-growing shrubs such as lavenders (*Lavandula*), heaths (*Erica*), and heathers (*Calluna*), as well as shaping larger shrubs, climbers, and small hedges. While excellent results can be obtained by using shears to prune a large hedge, it is hard and time-consuming work, so most people prefer electric or petrol-powered hedge trimmers to get the job done faster and with less effort. Before you buy, hire, or borrow hedge trimmers, always check that you can lift them comfortably, because they can be heavy.

Where precise trimming is required, smaller topiary shears are the best option. Their short handles are squeezed with one hand to bring the sharp blades together to cut through young shoots. This offers much more control than hedging shears.

Good-quality bypass secateurs are an essential tool in any garden.

Choose sharp loppers for cutting woody stems at the base of shrubs.

PROTECTIVE EQUIPMENT

Take measures to protect yourself, mostly from the thorny defences of plants and also from sharp tools, to avoid injury. Wear long sleeves and trousers to cover your skin and use thornproof gloves, made of thick leather, when working on plants with thorns or those that can irritate skin with their sap (*Euphorbia*) or hairs (*Fremontodendron*). Long leather gauntlets, which protect your arms up to the elbows, are invaluable when pruning large bush or climbing roses, or the long thorny stems of ornamental brambles (*Rubus*).

Eye protection is always advisable, because prunings may scatter and long stems spring back during training, while trimming wall-trained shrubs and hedges creates dust and debris. Powered hedge trimmers are noisy tools, so the use of high-quality ear defenders is recommended.

Protect your eyes and hearing when using powered hedge trimmers.

Sturdy gardening gloves are essential when working on plants armed with sharp thorns or those that are toxic.

SAFETY

You will do a better job and be less likely to strain or injure yourself if you use the right tools for each job and make sure that they are sharp before you begin. With electric power tools, always use a socket or adaptor with a RCD (residual current device) circuit breaker to protect you if you accidentally slice through the cable. Never use electric power tools in wet conditions.

If you need a ladder for a job, climb it only when it is on a firm stable surface. Never over-reach from it to make cuts, and avoid being on your own when carrying out work from a ladder.

If you are in any doubt about your ability to complete a job safely, bring in a qualified professional to do it and always use the services of a professional tree surgeon if a larger tree needs pruning. Such work can be dangerous if not handled by a skilled specialist.

NEED TO KNOW
- Clean, dry, and oil tools after use.
- Sharpen the blades of cutting tools regularly.
- Disinfect tools used to cut diseased material before using them on a different plant.
- Do not use secateurs to cut wire and string, as this will blunt them. Use wire cutters and scissors instead.

WHEN TO PRUNE

Correctly timed pruning helps to safeguard the health of plants and is also crucial to prevent the accidental removal of wood that will bear the next flush of flowers and fruit. You should therefore always check the best time to prune each of your plants by referring to its entry in the A–Z (see pp.40–141). You can then enjoy watching your garden flourish. Pruning methods also change according to a plant's age and stage of development, and this may affect the optimum time to prune.

IN SPRING

Many shrubs and climbers that flower in late summer or autumn on new growth are pruned in early spring, to promote flowering and to prevent plants becoming congested. All the stems are either cut to healthy buds just above a permanent woody base – as on bluebeard (Caryopteris) – or are pruned back to 30cm (12in) above ground level – as on fuchsia. Prune coppiced or pollarded shrubs and trees (see pp.28–29) from early to mid-spring. Keep low-growing, evergreen shrubs such as hebe compact by lightly trimming with shears in spring.

MORE PLANTS TO PRUNE IN SPRING
Butterfly bush (Buddleja davidii) • Cotton lavender (Santolina) • Lavatera • Perovskia • St John's wort (Hypericum)

Deadhead roses by cutting back stems just above a healthy bud.

Prune flowering currant (Ribes) after flowering to promote new growth.

AFTER FLOWERING

Shrubs or climbers that flower on stems produced and ripened in the previous growing season should be cut back as soon as possible after flowering. If you leave it too late to prune, you may forfeit next year's blooms, so be alert from early spring when tough shrubs like winter jasmine (Jasminum nudiflorum) need pruning, right through to midsummer, when you should have your secateurs ready as the fragrant blooms of mock orange (Philadelphus) fade. If you want a particular variety to produce ornamental berries or fruit in autumn you should refrain from pruning after flowering.

On plants that are not pruned after flowering, you should just deadhead (cut off the old blooms) wherever practical, to prevent them rotting and spreading disease. Pick off the flowers by hand, cut back to the nearest leaf bud, or trim over the whole plant lightly with shears.

MORE PLANTS TO PRUNE AFTER FLOWERING
Brachyglottis • Clematis montana • Deutzia • Forsythia • Lavender (Lavandula) • Weigela

Prune elder (Sambucus) in early spring to encourage young growth.

IN WINTER

Deciduous trees are dormant from late autumn through winter, and this is generally the best time to prune them. The bare leafless branches afford you a clear view of their structure, making it easy to assess where to cut, while access with tools is less awkward.

Work to renovate deciduous shrubs and climbers should also be carried out during late autumn and winter. Thin congested plants by removing old unproductive stems, or cut all growth right back to the base if appropriate (see the A–Z, pp.40–141). Plants will then produce plenty of healthy new growth in spring.

PLANTS TO PRUNE IN WINTER Beech (*Fagus*) • Climbing and rambling roses (*Rosa*) • Crab apple (*Malus*) • Hawthorn (*Crataegus*) • Wisteria

Prune most deciduous trees before their dormant buds burst open in spring.

> **TOP TIP** AVOID PRUNING EVERGREEN PLANTS BETWEEN EARLY AUTUMN AND EARLY SPRING, BECAUSE YOUNG SHOOTS AND RECENTLY CUT STEMS CAN BE BADLY DAMAGED BY FROSTS AND COLD WINTER WEATHER.

FORMATIVE PRUNING AFTER PLANTING

Pruning carried out to determine the shape of woody plants during their early years is known as "formative" pruning or training. This is the stage where important features such as a trunk clear of shoots and a framework of well-spaced branches are established in order to create an attractive plant when more mature. Thus, pruning at the right stage of development is vital.

Deciduous trees and shrubs tend to need more formative pruning than evergreens, while all climbers benefit considerably from initial training towards and against their support.

You can choose how much of this work to do by either buying a very young plant or else opting for a more mature, ready-pruned specimen, which will be more expensive.

MAINTENANCE PRUNING

Routine pruning helps to keep established plants healthy, vigorous, and looking their best. It should concentrate on dealing with dead, diseased, and damaged stems as soon as any are identified. After that, the amount and frequency of pruning required will vary from plant to plant.

Many trees and shrubs benefit from the removal of crossing, rubbing, or overcrowded branches, which increases the amount of light and airflow reaching their foliage. It is also good practice to cut out about one-third of the older stems at the base of some shrubs and climbers, to prevent congested growth and to encourage new shoots.

Plants allowed to grow in their natural form need less attention than those shaped into hedges or topiary, or trained against a wall, which always need pruning at least once a year.

Prune away congested branches to enhance a plant's health and appearance.

Remove outward-growing shoots from a shrub being trained against a fence.

HOW TO MAKE PRUNING CUTS

It is important to know exactly how to make pruning cuts on different plants, not only to encourage new shoots where you want them but also to allow healing to occur as quickly as possible. Therefore you need to take a careful look at your plants to see how the arrangement of buds along a stem can differ, because their patterns influence the way cuts should be made. Always use sharp tools to make smooth clean cuts and avoid any crushing or tearing, which injures stems and leaves the plant susceptible to disease.

HOW FAR FROM THE BUD?

The stems of all plants should always be cut just above a node, which is the point on a stem from which a leaf or shoot develops. Look for a node with a plump healthy bud. Recognizing such buds is usually easy as they protrude from stems at regular intervals and are often a darker colour than the bark on woody growth. Sometimes, dormant buds are flat and more difficult to identify, but can be found by looking for a spot-like mark on the stem with a curved line beneath it. Aim to cut about 3mm (⅛in) above the bud. If you cut too close, you may damage the bud; if too far away, you will leave an unattractive stub of wood that will die back and may allow disease to enter the plant. To ensure accurate pruning cuts, position the thin cutting blade of your secateurs closest to the bud. This may involve turning the secateurs over to align them correctly.

A correct pruning cut is one made just above a fat healthy bud.

When a stem is cut too far away from a bud, the resulting stub will die back.

Cutting too close to the bud risks damaging it and may prevent it developing.

DIRECTION OF GROWTH

By cutting to a bud pointing in the direction that you would like new growth to follow, pruning can also be used to shape a plant. To encourage an open structure on a freestanding tree or shrub, cut 3mm (⅛in) above an outward-facing bud (or buds); this will avoid a tangle of congested branches at the centre, which are likely to cross and rub together. Prune climbers and wall-trained shrubs just above any buds that will form new shoots to extend the branch framework in the desired direction or fill an area that is yet to be covered. Where buds have already developed into sideshoots, cut back in the same way, to just above the point where the shoot originates. Where only a single new shoot is needed from a pair of opposite buds, remove the unwanted bud by rubbing it off with your fingers.

Prune bush roses back to outward-facing buds, to create a cup-shaped form.

CUTTING STEMS WITH ALTERNATE BUDS

Buds borne first on one side of a stem and then the other in a repeating pattern are known as alternate buds. Prune plants with such a bud formation (for example, roses and wisteria) by making cuts that slant away from the bud, to allow water to run off the cut surface easily and so help reduce the risk of disease.

Use secateurs to begin the cut just above the base of the bud on the opposite side of the stem and angle the blades upwards so that the cut finishes 3mm (⅛in) above the bud itself, although it is advisable to look at each cut individually rather than worry about precise measurements. Use this technique also for single buds that spiral up a stem.

Make angled cuts sloping away from buds that are arranged alternately.

CUTTING STEMS WITH OPPOSITE BUDS

Some common garden plants such as buddleja and Mexican orange blossom (*Choisya*) produce buds in pairs, which are set exactly opposite each other along a stem. The correct way to prune such growth is to cut straight across the

Straight cuts just above the node allow opposite buds to grow away strongly.

stem above a pair of healthy buds. To avoid leaving a stub that will die back, make the cut as close to the buds as possible without damaging them. Take particular care when cutting above buds that have already formed shoots, because it is easy to damage the soft new growth when positioning secateurs near their base to make a cut. In a few plants, three or more leaves emerge from a single point on a stem, in what is known as a whorl. Prune above a whorl in the same way as you would opposite buds.

REMOVING A TREE BRANCH

Where it is necessary to prune a tree branch back to a larger main branch or the trunk itself, use a sharp pruning saw to cut just outside the branch collar – the slight swelling at the base of the branch. The collar is vital for proper healing once a branch is lost. First make

an incision underneath the branch just outside the collar, to about one-third of the way through the branch; this will prevent the branch tearing and leaving a jagged rip. Then do the main cut from the top, angling the saw slightly away from the trunk until it meets the undercut. Reduce the weight of moderately heavy branches by initially cutting them 30cm

(12in) from the trunk before removing the branch in the way already detailed. Never attempt to remove a branch that is too large to be cut with a folding pruning saw or a branch at a height that would require you to work from a ladder. Cutting large limbs can be hazardous and should be left to professional tree surgeons.

An initial undercut stops the bark tearing when a branch is removed.

Saw downwards, slightly away from the branch collar, to meet the undercut.

Never cut flush with the trunk as it reduces a tree's ability to heal effectively.

WHERE TO PRUNE

Once you have mastered how to make a pruning cut, you then need to work out what area of your plant should be pruned in order to produce the results that you are looking for. However, different plants have their own specific pruning requirements. By gaining an understanding of the general techniques needed to remove unwanted stems, as well as to encourage dense growth and to promote flowering and vigorous new shoots, you will be able to choose an appropriate cutting tool and use it confidently, without the need for a reference book by your side.

Remove "reverted" green shoots from variegated plants by cutting them right back to their point of origin.

Pinch out shoot tips of wall-trained shrubs to produce bushy growth.

AT THE SHOOT TIP

Tip-pruning involves pinching out or cutting back the top of a shoot, often just to the nearest healthy bud, to stop extension growth and to stimulate sideshoots. Tip-pruning is most frequently employed to keep shrubs bushy and climbing plants within their allotted space. It can also be used to shape young plants during their formative pruning after planting, and also to encourage overly vigorous stems to branch lower down. Unproductive blind buds, faded flowers, and any damage caused by frost can be removed at the same time.

AT THE BASE OF A SHOOT

The best way to control unwanted growth is to completely remove shoots by cutting them back flush against the stem on which they originate. No dormant buds can then regrow. This technique is especially helpful when training plants into a specific form, such as espalier (see p.39), because it allows you to cut away stems growing in the wrong direction and, if appropriate, to remove shoots in order to create a clear trunk. Other reasons to prune at the base of a shoot include removing any "reverted" shoots, which bear all-green leaves on what should be a variegated plant. The long vertical shoots that arise near pruning cuts on tree trunks or branches, known as "water shoots", are also undesirable. Prune water shoots right back to their point of origin as soon as they are spotted, or rub away the buds that will form them.

Water shoots, which arise from around old pruning wounds, should all be removed completely at their base.

Prune the previous season's sideshoots back to two or three buds.

ON A SIDESHOOT

Shortening sideshoots to stimulate flower production, and the subsequent formation of fruits, is known as spur-pruning. This technique is most commonly used to encourage a dense form and prolific display of blooms on shrubs such as Japanese quince (*Chaenomeles*), climbers such as wisteria, and a few trees such as laburnum that are trained against or over supports.

When spur-pruning, sideshoots are cut back to within two or three buds of the lateral stem from which they arise, diverting the plant's energy into the formation of flower buds, rather than the longer leafy stems that would naturally otherwise be produced.

NEAR THE PLANT BASE

By removing a proportion of an established plant's stems every year, you help to prevent it from becoming congested. Pruning in this way also boosts the formation of young stems. This technique is most helpful where stems emerge at ground level on shrubs such as deutzia and mock orange (*Philadelphus*), but can also be used to remove tired old stems from many shrubs and climbing plants to encourage strong new growth.

Select between one-fifth and one-third of the oldest stems, which can be recognized by their different colour and bark texture compared to younger shoots. Cut each stem either at its base or at 5–8cm (2–3in) above ground level. Use loppers or a pruning saw to cut cleanly through stems thicker than 1cm (½in) in diameter.

Remove old mock orange stems at the plant base to stimulate new shoots.

REMOVING SUCKERS

Watch out for upright shoots, known as suckers, arising from the roots of plants or the base of hybrid roses where your chosen variety has been grafted on to a more vigorous rootstock. Suckers grow rapidly and will weaken or dominate plants if left unchecked. When you spot a sucker, immediately remove it by pushing aside the soil to find its point of origin and pulling the sucker away so that any dormant buds around it are also included. Never cut a sucker, because it will then quickly regrow.

Rose suckers have distinctive foliage and should be pulled away firmly.

BACK TO A PERMANENT WOODY FRAMEWORK

Plants that flower in late summer or autumn on their current season's growth, such as butterfly bush (*Buddleja davidii*) and bluebeard (*Caryopteris*), and those that are coppiced or pollarded for young foliage or brightly coloured stems (*see pp.28–29*), are cut back close to a permanent woody base, stem, or framework of branches. This hard pruning stimulates vigorous new growth.

Prune just above a fat healthy bud or pair of buds where they are visible (*see pp.20–21*), leaving just two or three buds above the framework of older growth. Sometimes, dormant buds at the base of stems may be difficult to spot. Where it is not possible to see buds, cut stems in the required position and, once growth is underway, remove any significant stubs above the new shoots to improve the plant's appearance and prevent dieback.

Butterfly bush quickly forms shoots when pruned back to a woody base.

GENERAL TRIMMING

When shoot tips are evenly removed over the entire surface of a plant it is known as trimming or clipping. This shaping technique is invaluable for encouraging bushy growth in a select range of plants with compact habits or dense foliage, or those trained as topiary or hedges. However, it should be carefully timed and used only where recommended (*see pp.40–141*), to avoid losing the flowers and natural form for which many plants are prized. Trimming can be carried out with secateurs on small plants, but use hedging or topiary shears – or hedge trimmers – for larger shrubs and hedges.

SHAPING TREES AND SHRUBS

Trimming with hedging shears can be a useful way to enhance the natural forms of densely branched, evergreen trees such as holly (*Ilex*) and bay (*Laurus*). Their tightly packed foliage quickly creates a well-defined shape when trimmed annually, which helps to keep these large plants within bounds in a garden setting and adds useful winter structure. Few other trees and large shrubs will benefit from regular trimming and if you find yourself wanting to use shears to limit the size of these plants you should first consider whether they still have ornamental value or if it would be better to replace them.

For many small and low-growing shrubs, however, trimming is the ideal way to prevent spreading leggy growth, and bareness in the centre. Compact shrubs such as heathers (*Calluna*), hebes, and santolina are trimmed after flowering, which removes both the faded flowerheads and the shoot tips so instantly tidies the plants and promotes plenty of bushy new growth.

TRIMMING CLIMBERS

Many rampant climbers such as honeysuckle (*Lonicera*), *Clematis montana*, and *C. tangutica* cover an extensive area with a mass of spindly stems and do not need any special pruning to promote flowering. Trimming them lightly with shears each year, after their blooms have faded, keeps them tidy and confined to their allotted space and supports.

Robust climbers that are grown for their foliage, such as Virginia creeper (*Parthenocissus*), should be trimmed while dormant in winter if they are deciduous or in early spring before growth begins if they are evergreen. It is not essential to trim every year, but regular cutting helps to prevent plants becoming bare at the base and tangled in areas where they may cause damage.

Keep lavender (*Lavandula*) neat by cutting its faded flowerheads with shears.

Cotton lavender (*Santolina*) is dense and mounded when clipped annually.

Trim Virginia creeper (*Parthenocissus*) once its fiery autumn leaves have fallen.

CLIPPING TOPIARY

Topiary can raise the training and clipping of small-leaved evergreens such as box (*Buxus*) to an art form. Maintaining the crisp lines of a striking geometric shape or an almost unlimited array of imaginative forms requires patience, precision, and sharp cutting tools. Shears can be used for large areas, while secateurs or topiary shears allow finer control, which makes them better for crafting intricate details. Always clip with the flat edges of blades parallel to the surface being cut. Remember to step back regularly to ensure that your desired shape is retained, so you do not focus on one area.

The frequency of trimming depends on the species used and how crisp you need your topiary to look. Complex designs and prominent geometric forms may require attention every six weeks or so. Clipping twice during the growing season (in late spring and late summer) is usually sufficient for box topiary, while more vigorous species (for example, *Lonicera ligustrina* var. *yunnanensis*) need to be clipped three times a year to stay neat.

Never prune topiary plants after early autumn, to allow new growth to ripen so that it can withstand winter weather.

Hedging shears are ideal for clipping the angled shape of a box cone.

Regular and precise clipping has created the elegant curves and dense foliage on this *Lonicera ligustrina* 'Elegant' hedge.

PINCH-PRUNING

By nipping off the growing points over the whole or the majority of a plant between your thumb and forefinger, you promote the growth of sideshoots in exactly the same way as trimming with shears. This method is known as pinch-pruning, while tip-pruning (see p.22) is carried out on just a small number of a plant's growth points.

Use pinch-pruning in spring and summer to create a bushy-shaped plant, increase the number of flowering stems, and sometimes deliberately delay flowering in herbaceous perennials and subshrubs such as fuchsia. It is also the technique employed to form the dense head of foliage that tops the clear trunk of plants trained as standards (see p.30). Newly planted shrubs with an uneven shape also benefit from pinch-pruning.

Pinch off the tips on sage (*Salvia*) to keep its growth compact and even.

TOP TIP MANY SPRING- AND EARLY SUMMER-FLOWERING HERBACEOUS PERENNIALS SUCH AS PULMONARIA AND HARDY CRANESBILL (*GERANIUM*) CAN BE TRIMMED BACK HARD AFTER FLOWERING TO PRODUCE A FRESH FLUSH OF FOLIAGE AND SOMETIMES FLOWERS.

CLIPPING HEDGES

Hedges are prominent garden features which create effective windbreaks, desirable privacy, and attractive backdrops for other planting. They are formed by planting shrubs or trees at close intervals, then pruning and training their growth to the desired shape. Regular, well-timed trimming is essential to keep hedges dense and to maintain their intended appearance, whether that's the crisp edges of a formal hedge or a covering of colourful summer blooms for a more informal style. Keep hedges in good health by applying a balanced fertilizer and mulching each spring.

Electric hedge trimmers produce a neat finish when used with due care.

Guide hedging shears along a string line for a perfect level top.

PRUNING A NEW HEDGE

The way that young plants are pruned after planting governs a hedge's eventual height and width and also encourages consistently thick growth.

Conifers and most evergreens can be stimulated to thicken by gently trimming their lateral shoots during summer. Do not cut their leading shoots until they have reached the desired height.

Deciduous hedges planted with stocky species such as hornbeam (*Carpinus*) and beech (*Fagus*) should have both their leading shoot and lateral shoots pruned back by roughly one-third during late winter after planting, and again in their second year.

Species with more upright growth, such as hawthorn (*Crataegus*) and privet (*Ligustrum*), need tougher treatment to promote bushy growth right from the base of the plant. Cut them down hard to 15–30cm (6–12in) above ground level in the late winter or early spring after planting, trim lateral shoots in summer, and shorten new growth by one-half at the end of the next winter.

ESTABLISHED HEDGES

Check hedges for nesting birds in spring or summer and, where necessary, delay trimming until nests are no longer in use. Sharp hedging shears or hedge trimmers are used to clip most hedges. Their long flat blades make it easy to produce straight sides and a level top, and they allow you to cut large areas quickly and efficiently. Keep the blades of your cutting tool parallel to the surface to achieve a straight cut and avoid biting into the hedge and spoiling its outline. Take regular breaks to avoid fatigue and step back to check your progress. It can be helpful, particularly when trimming long hedges, to use a string line pulled taut along the top of the hedge between two posts as a guide to produce a level result.

Trim hedges formed from large-leaved shrubs, such as *Aucuba japonica*, with secateurs, to avoid making unattractive cuts across the foliage, which will turn brown at the edges.

A newly planted hedge requires careful pruning in its early years to stimulate dense foliage within your desired height.

FORMAL HEDGES

Formal hedges are often cut with gently slanting sides so that they are slightly wider at the bottom than the top. This is known as a "batter" and helps snow to slide off the foliage as well as allowing light in to the base, promoting good growth. The clearly defined outline of a handsome formal hedge is always the result of careful clipping carried out at the right time of year. It is not chance.

To achieve the dense growth necessary for a really neat appearance, most hedges need to be trimmed twice a year. Deciduous hedges, such as beech (*Fagus*), should be cut up to three times a year. Trim conifer hedges and those formed from broad-leaved evergreens, such as holly (*Ilex*), once the weather is mild in late spring, and then no later than late summer, to give new growth time to toughen up before winter, when it is vulnerable to frost damage. It is essential to prune conifer hedges regularly to keep their growth dense and their size manageable, because most will not produce shoots if cut back hard into older wood.

A crisply trimmed yew (*Taxus*) hedge is the perfect backdrop for features like colourful planting or this white bench.

INFORMAL HEDGES

Not all hedges need to be uniform and neatly clipped. Less rigorous pruning allows flowering and berry-bearing shrubs to be grown as informal hedges, creating a colourful display. They can also be a great resource for wildlife. Cotoneasters, fuchsias, firethorn (*Pyracantha*), and forsythias will all form attractive hedges, but it is vital to time their pruning as recommended in the A–Z (see pp.40–141), to avoid cutting off the shoots that will produce flowers and fruits. Mixed hedges can be created by planting several species of native trees and shrubs together.

Flowering hedges often suit gardens with a more naturalistic planting style.

PRUNING FOR SPECIFIC EFFECTS

Coppicing and pollarding are simple traditional pruning methods, frequently used in gardens to transform the habit and ornamental qualities of trees and shrubs that tolerate being cut back severely. This repeated hard pruning produces flushes of vigorous young shoots, which are valued for their vibrantly coloured bark or unusually large leaves. These multiple unbranched young shoots are quite different from a plant's natural form and can be a useful way to fit trees and shrubs into a small garden or among herbaceous plants without casting too much shade.

COPPICING

This involves cutting all growth back close to ground level to produce, and then regularly renew, a crop of straight young stems. Not all trees and shrubs respond well to such drastic treatment; use only plants for which coppicing is recommended in the A–Z (see pp.40–141). Many plants are coppiced annually, because it is the youngest stems that produce the most striking visual effects, but others such as hazel (Corylus) look their best when cut back roughly every five years.

Allow new plants that are to be coppiced to grow unpruned for two or three years after planting, so that they are well established and able to recover quickly from drastic treatment. Pruning is usually carried out while plants are dormant, in late winter or early spring. All stems are cut back to a uniform height, as little as 8cm (3in) above the ground level. After coppicing, support the subsequent vigorous growth by applying fertilizer and a thick mulch of well-rotted compost to the surrounding soil. Healthy plants will continue to respond well to coppicing for many years.

MORE TREES AND SHRUBS FOR COPPICING AND POLLARDING Dogwood (Cornus) • Elder (Sambucus) • Foxglove tree (Paulownia) • Gum (Eucalyptus) • Indian bean tree (Catalpa) • Judas tree (Cercis) • Populus × jackii 'Aurora' • Smoke bush (Cotinus) • Willow (Salix)

> **TOP TIP** USE SHARP LOPPERS OR A PRUNING SAW TO CUT CLEANLY THROUGH THE THICK STEMS AT THE BASE OF PLANTS AND SO REDUCE THE CHANCE OF DISEASE WHERE PLANTS ARE COPPICED OR POLLARDED FREQUENTLY.

Uniform pruning of hazel close to ground level forms a coppice "stool".

Erect new growth is produced from the outer edges of the hazel stool.

Coppiced willow has a shrub-like form and provides supple stems for weaving.

Pollard the bright stems of *Salix alba* var. *vitellina* 'Britzensis' in early spring.

The vibrant stems of coppiced dogwood and willow provide a colourful winter display even through the toughest weather.

POLLARDING

The same profusion of young stems as coppicing is achieved by pollarding, but they are dramatically displayed at eye level on top of a single woody trunk or branch framework. Large trees are often pollarded to reduce their size, while gums (*Eucalyptus*) and willows (*Salix*) are among a number of trees and shrubs that respond well to the same treatment in small gardens.

To create a pollard on a single trunk, plant a young tree that has a good head of branches above a clear stem about 2m (6½ft) tall. Allow the tree to establish for three years. Then, before the diameter of the main stem exceeds 13cm (5in), cut it just above the lowest branches in winter or early spring. Shorten these branches back to about 2.5cm (1in) from the main stem; new shoots produced from buds below these cuts will then form a decorative head of young growth.

Thereafter, in late winter or early spring every one to three years, prune back close to this framework. At the same time, thin out any overcrowded shoots and remove any growth that arises from the clear trunk.

COLOURFUL WINTER STEMS

When rising from the middle or back of a border, the fiery-coloured, young stems of coppiced or pollarded dogwoods and willows look striking when illuminated by low winter sunlight, and will also provide the perfect backdrop for flowering plants during summer with their attractive foliage. Stems range in colour from the bold coral-pink of *Salix alba* var. *vitellina* 'Britzensis' to the bright yellow-green of *Cornus sericea* 'Flaviramea'. The bare stems of such willows and dogwoods are pruned back at the last opportunity – as late as mid-spring – so that the display can be appreciated for as long as possible.

PRUNING FOR ORNAMENTAL FOLIAGE

When stimulated to produce young growth by coppicing or pollarding, some trees and shrubs assume an entirely different character by forming dense stands of lush leaves, which are larger and often more boldly coloured than the foliage of unpruned plants. Many gums (*Eucalyptus*), for example, respond to hard pruning by sending up stems of silvery, disc-shaped juvenile foliage, which are unlike the elongated leaves on mature plants. Some plants that are pruned to produce larger leaves, such as elder (*Sambucus*) and Indian bean tree (*Catalpa*), will not flower as they do this only on mature growth.

Huge, heart-shaped leaves have grown on this pollarded Indian bean tree.

PRUNING FOR DIFFERENT SHAPES

Pruning can be used to modify the shapes of trees and shrubs in a variety of ways. Striking formal effects can be achieved by pruning to concentrate stems, foliage, and flowers at the top of branch-free trunks on standards and pleached trees. Creating multi-stemmed trees is a great way to emphasize beautiful bark and to evoke the atmosphere of a woodland in a small space, while lifting the canopy of trees can improve the shape of more mature specimens and prevent them dominating your garden.

CREATING A STANDARD FUCHSIA

A shrub that has been pruned and trained into a "lollipop" shape, with a bushy head of lateral shoots and foliage on top of a tall clear main stem, is known as a "standard". Pruning in this way creates a formal effect, and such standard plants look particularly good when grown in smart containers. Fuchsia and bay (*Laurus*) are commonly grown as standards using young plants, but this technique does not suit all species, so check the A–Z (*see pp.40–141*) for those that respond well.

To create a standard fuchsia, begin by pinching out any lateral shoots that develop from the main stem when a young plant reaches 15cm (6in) tall.

Continue to pinch out all lateral shoots so the plant has a clear, lateral-free stem, and tie the main stem to a vertical cane as it gains height. Once the fuchsia has grown to just above your required height of clear stem, pinch out the growing tip to encourage the production of lateral shoots, which will form the head of the standard. Allow these lateral shoots to develop. Then, pinch out the top set of leaves on each lateral shoot, to stimulate bushy growth. Do this with your fingers on soft growth, but use secateurs if necessary. Continue to pinch out tips of further lateral shoots as they form, until you have a spherical head of stems and foliage. Once the desired shape is reached, stop pinch-pruning, to allow flowering. Leaves usually fall from stems naturally, but remove any that remain, along with any lateral shoots on the stem.

A standard fuchsia offers a spectacular display of flowers close to eye level.

PLEACHING

This traditional technique involves weaving together the branches of trees planted closely in a row to form what looks like a hedge raised up on straight clear trunks. It has now become a popular formal feature in contemporary gardens. Beech (*Fagus*), lime (*Tilia*), and hornbeam (*Carpinus*) all look spectacular when pleached and are ideal for providing privacy in overlooked gardens.

Pleached copper beech forms a screen without totally enclosing the garden space.

Plant during the dormant season, add a stake for each trunk, and leave the framework of supports for the branches in place. Weave and tie in suitably placed shoots along these horizontal supports in summer to thicken the screen. In winter, cut back shoots growing in the wrong direction and shorten long shoots to promote branching. Once a dense screen has been formed, growth should be trimmed twice a year like a hedge.

Because a pleached hedge is technical and time-consuming to create, most gardeners opt for the instant impact of ready-trained trees, despite the cost.

PRUNING FOR A MULTI-STEMMED FORM

Trees with several trunks can occur naturally, but single-stemmed, deciduous trees such as birch (*Betula*), cherry (*Prunus*), and maple (*Acer*) can also be pruned to create this pleasing effect, either on a low trunk (see p.45) or close to ground level.

Prune a young tree in winter by making a straight cut across its main stem at the desired height (at least 8cm/3in above ground level). The following winter, choose three or four well-spaced shoots of even thickness to form the main stems, and cut all other shoots back to the base. Thereafter, allow the tree to develop naturally, removing only shoots at the base in addition to lateral shoots that appear low on the main stems.

Never use this multi-stemmed pruning method on grafted trees, because you will cut off your chosen variety and be left with the rootstock.

The russet bark of *Prunus serrula* is accentuated in a multi-stemmed form.

Several multi-stemmed trees planted adjacent create a shady glade.

Use sharp loppers to prune any tree branch more than 1cm (½in) thick.

While sawing, support the branch to prevent its bark tearing at the base.

CROWN LIFTING

This technique, which is best reserved for deciduous trees, is used to increase the height of the branch-free, clear trunk below the lateral branches. It is an ideal way to create more space and allow in light beneath a tree, as well as to improve its shape and make an eye-catching feature of any colourful or textured bark. You should lift the crown of only small garden trees; such work on larger trees can be dangerous and should be carried out only by a professional tree surgeon.

Crown lift your tree in winter, when dormant. Avoid damaging species that bleed sap from pruning cuts, by timing their pruning carefully in early winter. Use loppers to cut thin lateral branches back to the raised "collar" where they meet the trunk, and switch to a pruning saw for thicker branches.

RENOVATION PRUNING

If your plants outgrow their space, become reluctant to flower, or just look tired, it may be possible to rejuvenate them using renovation pruning. Many healthy plants produce an astonishing response to severe pruning, and it can be a great way to save an old favourite. Drastic renovation will prevent flowering for a year or two, but the resulting young growth should be well worth the sacrifice. This treatment will not resurrect those trees, shrubs, or climbers weakened by disease or old age, however, and sometimes it is best to start again with a young plant.

Remove a proportion of thick tired stems each year to rejuvenate leycesteria.

RADICAL RENOVATION OF SHRUBS

Large overgrown shrubs can often be renovated by cutting all their growth down close to ground level in one go, to stimulate the production of vigorous new shoots. This can be particularly helpful for shrubs with thorny stems or foliage, such as holly (*Ilex*), which would be awkward to renovate gradually. Not all shrubs recover from drastic pruning, so always check the A–Z (*see pp.40–141*) before making cuts.

Renovate deciduous shrubs when dormant, between late autumn and early spring, and tackle evergreens in spring as growth begins. Use loppers or a pruning saw to cut straight across the main stems, usually 30–60cm (12–24in) above ground level. Remove lateral stems first to improve access and cut the main stems of very large shrubs about 30cm (12in) above the final pruning height, to reduce their weight and help ensure a clean final cut. The following winter, thin new growth around each cut, to select the strongest shoots and help create an attractive new framework of branches.

SOME SHRUBS SUITABLE FOR RENOVATION PRUNING Barberry
(*Berberis*) • Beauty berry (*Callicarpa*) • Brachyglottis • Deutzia • Fuchsia • Hawthorn (*Crataegus*) • Kerria • Lilac (*Syringa*) • St John's wort (*Hypericum*)

The prolific new growth on this renovated lilac will need to be thinned.

GRADUAL RENOVATION OF SHRUBS

Renovation pruning can also be carried out in stages over several years, while deciduous shrubs are dormant or in spring for evergreens. This option is kinder to plants that resent drastic pruning, such as silk tassel bush (*Garrya*) and Japanese quince (*Chaenomeles*), and also retains foliage and flowers during the process so that there is no bare gap in the garden. Start by removing dead, diseased, and damaged growth, then prune up to one-half of the stems – choosing the oldest and weakest first – back to 5–8cm (2–3in) above ground level or to a main framework of branches. To create a more compact form, shorten the remaining stems by about one-half of their height, to an outward-facing bud. During winter for deciduous shrubs or spring for evergreens, thin out the new shoots to produce a pleasing, uncluttered new shape.

Prune in the same way in the second year, cutting back one-half of the remaining old stems to the base or branch framework. In the third year, remove the last of the old stems.

REINVIGORATING CLIMBERS

Where old climbers are overgrown, or it becomes necessary to repair their supporting structure, it is often best to cut the whole plant down to 30–60cm (12–24in) above ground level. This quickly removes the tangle at the top of the plant and will encourage vigorous new shoots, which can be trained on to supports to create the desired effect.

Many deciduous climbers including honeysuckle (*Lonicera*) and passion flower (*Passiflora*) tolerate drastic pruning when still dormant in early spring, but consult the A–Z (see pp.40–141) before undertaking any work. Renovation over two or three years is a better option for climbers that don't respond well to drastic pruning, for example, hydrangea vine (*Schizophragma*), or where you don't want to sacrifice all foliage cover or flowers in the short term. However, it is more labour intensive, because when only two or three of the oldest and weakest stems are cut out each year their growth needs to be carefully disentangled from the supports.

After cutting back an old *Clematis montana*, pull the knotted stems from its support to make way for new shoots.

Renovate climbing roses gradually by cutting back the oldest stems to the base when they lose their vigour.

RENOVATING HEDGES

Even with regular trimming, hedges can eventually creep outwards or upwards over the years to become too wide or too tall for their space. Many deciduous hedging species such as beech (*Fagus*) can be brought back within bounds using hard renovation pruning in midwinter, but of the conifers commonly used for hedging only yew (*Taxus*) will regrow when cut back hard to old wood in mid-spring.

If severe pruning is necessary, renovate one side by cutting growth back uniformly, close to the main stem in the first year and, if new growth has been vigorous, prune the second side in the same way the following year.

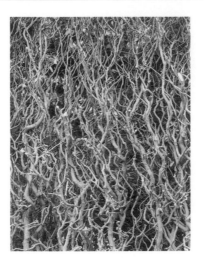

Reduce a hedge's width by pruning one side back hard each year.

Yew hedges will quickly send out a dense mass of young shoots after pruning.

TRAINING PLANTS

You can produce some spectacular effects in your garden, such as a flowering screen along a boundary fence or a silk tassel bush (*Garrya*) shaped into an elegant espalier, by using various training techniques. These manipulate a plant's natural growth habit, which causes it to develop a different shape or to enhance desirable qualities such as flowering or fruiting. Most effects are easy to achieve, and the more complicated methods are also not difficult to master with practice. Beginners may wish to start with training a climber to cover a wall or fence.

The rambling rose 'Paul's Himalayan Musk' has been trained up a mature tree to decorate it with early summer blooms.

WHY TRAIN PLANTS?

One of the main reasons for training a plant is aesthetic, although it also offers similar benefits to general pruning, such as stimulating growth, encouraging flowering and fruiting, and keeping plants healthy. Consider a rose arch covered with blooms or an elevated, box-shaped screen created by pleached trees (see p.30) providing extra privacy in a town garden. Both of these effects are achieved by the training method of tying stems to a support structure, pruning to create the appropriate shape, and then maintaining it once or twice a year.

Training is also useful in limiting the size of a tree or other type of plant. Trimming a conifer such as yew (*Taxus*) to create topiary or training a shrub against a wall will produce compact, eye-catching features for a small garden.

Some flexible-stemmed plants are trained to form freestanding features. For example, supple willow (*Salix*) stems harvested from pollarded trees (see p.29) will root when inserted into soil and can then be woven together to create living arches and tunnels.

Honeysuckle (*Lonicera*), clematis, and roses have been trained over this arch.

WHICH PLANTS CAN BE TRAINED?

Most climbing plants such as roses and wisteria are trained on to a support to prevent their towering stems from growing skyward and their flowers blooming in clusters at the top, where they may be too far up to be appreciated. Shrubs with flexible stems such as ceanothus and Japanese quince (*Chaenomeles*) can also be trained on wires or trellis to cover a wall or fence, and a bay tree (*Laurus nobilis*) will make an attractive feature when clipped into a "lollipop" standard.

Bay trees are easily trained to form standards by removing the lower stems.

Pyracantha stems trained on wires fixed to a wall will encourage a profusion of decorative autumn berries to form.

EASY PLANTS TO TRAIN

Climbers are among the simplest plants to train because their flexible stems are designed to twine or cling to a support in a variety of ways (see below). The twiners, such as clematis and honeysuckle (*Lonicera*), will just require the right type of support (see pp.36–37) and a little encouragement in order to cover a wall or pergola with their decorative flowers and foliage. Climbing roses should be tied on to their individual supports and the stems trained horizontally to achieve the same wallpapering effect.

Clematis stems should naturally twine around a narrow-gauge trellis to cover it.

INCREASING FLOWERING AND FRUITING

The growth of lower buds or shoots on a stem, which will subsequently bear flowers and fruits, can be stimulated by training. The technique relies on removing or disrupting the hormones flowing to the dominant buds on the leading stem or stems (see *Apical dominance, pp.14–15*). When these stems are tied almost horizontally, the plant responds by making more lateral stems and sideshoots, and these in turn produce an abundance of flowers followed by fruits, which may be ornamental, such as rose hips, or edible, in the case of apples and pears. Once you have created the basic framework for your plants and carried out the initial training, this task will become more routine year by year as you master the technique.

HOW DO CLIMBERS CLIMB?

Climbers use a range of different techniques to climb. Some are self-clinging, while others will need assistance to grow up a support.

ADHESIVE ROOTS AND PADS
Climbers such as ivy (*Hedera*), Virginia creeper (*Parthenocissus*), and climbing hydrangeas (*H. petiolaris*) cling with adhesive aerial roots or pads that stick to surfaces without the need for wires, canes, or trellis. Many are vigorous and will outgrow their allotted space if they are not cut back regularly.

LEAF-STALK AND TENDRIL TWINERS
The most well-known of the twiners is clematis, which uses its leaf stalks to curl around trellis, canes, or wires. These climbers require slim supports – about the width of a bamboo cane is ideal – to twirl around. Jasmine (*Jasminum*) and vines (*Vitis*) use leaf tendrils to climb, and also require narrow-gauge supports such as canes or wires to climb up.

STEM TWINERS Wisteria, honeysuckle (*Lonicera*), and false jasmine (*Trachelospermum jasminoides*) gain height by using their stems to twine

around supports. In addition to slim supports such as wires, most will encircle wider structures than the leaf-stalk and tendril twiners – wooden trellis is ideal. Wisteria will need a sturdy frame to support its weight when established.

THORNY HOOKS
Climbing and rambling roses use their backward-curving thorns to hook on to natural supports such as trees. Tie such stems to wooden posts or trellis, which the hooks cannot grip.

SUPPORTS FOR PLANTS

Choosing the right type of support for a climber or wall plant is critical if you want to train it into a beautiful flowering or fruiting feature and prevent it from growing into an untidy tangle of stems. Always ensure that the support you create or buy will be large and strong enough to accommodate your plant, and fix wires or trellis securely to provide a sturdy framework that will hold the plant as it matures.

Climbing roses and clematis can be trained on wires attached to posts.

WALLS AND FENCES

To brighten up your boundaries with a festival of flowers, foliage, berries, and seedheads, train your climbers and shrubs against a wall or fence. Climbers with adhesive roots and pads will grow up these surfaces unaided, while those that use other methods to climb (see p.35) will require some assistance to encourage them to do so, as will wall shrubs. Plastic-coated or galvanized horizontal wires fixed every 30cm (12in) up a wall or fence are ideal for most plants. When using wires, you may also need a few canes angled towards the bottom of the support structure. These will allow twining plants to make their initial ascent. As they grow, their long stems should reach up to the next set of wires of their own accord.

Many climbers, including twiners and roses, will also climb up trellis. Buy a product that is the right height and width to cover your wall or fence and suitable for the plant you have chosen (see pp.34–35). Plant your climber at least 45cm (18in) from the structure.

Secure trellis to wooden battens screwed to a wall or fence.

ATTACHING TRELLIS TO A WALL

To allow air to circulate freely and plant stems to twine through trellis, there should be a short distance between a plant and the wall or fence. Measure and cut two wooden battens to the same width as the trellis panel. Using stainless steel screws and an electric screwdriver, secure one batten horizontally along the top of the wall or fence. Measure the height of the trellis panel. Mark this distance from the top of the first batten to where the trellis will fit on wall below. Secure the second batten horizontally at this point, so that the trellis fits neatly on top of the two battens. Finally, screw the trellis firmly to the battens.

Japanese quince (Chaenomeles) has flexible stems that can be trained along horizontal wires to produce a flowering wall.

FREESTANDING STRUCTURES

When training plants on to freestanding structures such as pergolas and arches, you may need to attach wires to the vertical posts to enable them to climb (see right). Alternatively, opt for a metal-framed pergola with slim

Clematis stems may need to be tied on to wooden tripods with wide posts.

supports that a twiner can cling to more easily. Plant climbers at least 45cm (18in) from the vertical supports. Once at the top, their stems can be attached to the roof beams with string or plant ties or, in the case of clematis, honeysuckle (*Lonicera*), and jasmine (*Jasminum*), left to scramble over the supports. All plants grown over such freestanding structures require annual pruning (see the A–Z, pp.40–141).

Small stem-twining climbers can be grown up obelisks and tripods without additional wires, but if you want to grow tendril or leaf-stalk twiners (see p.35) the timbers or other supports should be no more than 1.5cm (½in) wide.

Growing climbers up into trees and large shrubs is relatively easy, although you initially may need some canes or a strip of small-gauge trellis to give the stems a framework on which to climb up to the lowest branches. Thereafter, they will scramble through the tree or shrub to create a naturalistic effect. Rambling roses make a beautiful feature growing through a tree, but only try this if the tree is mature enough to take the rose's weight. Ideally, plant about 2m (6½ft) away from the tree trunk.

TYING IN

Only the young stems of woody climbers, such as roses and wisteria, are flexible enough to be bent horizontally towards a support and secured in position. If stiff older wood requires training, prune the stem just below a wire and tie in the supple new growth as it emerges.

To avoid damaging a plant's stems use soft twine or proprietary plant ties. With twine, make a figure-of-eight around the plant and its support. Tree ties may be a better option for heavier plants, such as climbing roses or wisteria. Check ties regularly and replace as the stems expand to ensure they do not restrict growth.

Tie stems to supports with soft twine strung in a figure-of-eight.

Fix screw eyes to the top and bottom on each side of the post.

Thread wires through the screw eyes and tie your climber to the wires.

FIXING WIRES TO A STRUCTURE

Wires and screw eyes are needed to hold plant stems evenly over a pergola or arch. Fix a screw eye to the top and bottom of each post on all four sides. Thread a length of wire through the screw eye at the top and twist to secure it. Pass the loose end through the eye directly below, pull it taut, and secure it. Repeat on all post legs.

Use the same technique to attach wires to a wall or fence. Space the screw eyes 45–60cm (18–24in) apart up each side of the fence panel or wall and then secure the wires horizontally.

TRAINING TECHNIQUES

Training a plant to take on a particular shape involves more than simply tying stems to supports. Before you even plant your climber or wall shrub, you need to check that its support is strong and in good condition and secured well into the ground or against a sturdy wall or fence. Protect any timber with wood preservative. Several coats may be required but wait until each is dry before applying more.

Climbing roses such as *Rosa* 'Zéphirine Drouhin' can be trained horizontally on sturdy wires to cover a wall or fence with abundant flowers.

TRAINING A CLIMBER
ON HORIZONTAL WIRES

In autumn or early spring, fix sturdy wires through vine eyes every 30cm (12in) up a wall or fence and tie bamboo canes on them to form a fan shape over the surface. Plant your climbing rose or other climber 45cm (18in) away from the wall and tie the main stems to the canes with twine, using a figure-of-eight knot. Then in summer, after removing any dead and diseased stems, and those that are rubbing others, cut the main stems back by 5–8cm (2–3in) to an outward-facing bud, using an angled cut. This will encourage lateral stems to form.

Each autumn thereafter, attach the young flexible laterals that have grown during the summer to the horizontal wires. Where a stem is crossing or rubbing against others, remove or reposition it to fill a space on the wires. Prune back lateral or sideshoots that have flowered by two-thirds of their length, making an angled cut just above a healthy bud or shoot. Cut old stems that have not flowered back to a healthy bud about 30cm (12in) from the ground. When a new shoot develops from this cut, tie it on to the wires to fill a space. Also shorten any stems that have outgrown their allotted space.

TOP TIP TRAIN RAMBLING ROSES HORIZONTALLY IN THE SAME WAY AS CLIMBERS BUT PRUNE THEM BACK AFTER FLOWERING IN SUMMER. ALSO MAKE SURE YOU HAVE A LONG ENOUGH FENCE OR WALL TO ACCOMMODATE THEIR VIGOROUS GROWTH.

Pyracantha espaliers create a spectacular effect against a wall.

CREATING AN ESPALIER

Training a plant to form an espalier is an advanced technique and beginners may prefer to buy a pre-trained form and then maintain it. More confident gardeners can buy a young plant such as such as pyracantha, Japanese quince (*Chaenomeles*) or silk tassel (*Garrya*) and start from scratch. Both young and pre-trained shrubs should be planted at least 45cm (18in) away from a wired-up wall or fence (see p.36).

To train a young plant, in spring, cut the main leading stem to a healthy bud just above the first horizontal wire. This will stimulate lateral shoots to grow just below it. After the plant has flowered (see the A–Z, pp.40–141), tie the leader and two opposite laterals to canes and attach the canes to the horizontal wires. Do not force the laterals down – fix them at 45 degrees to the ground at this stage and lower them to a horizontal position later in summer. Prune any other laterals back to two or three leaves.

During the following spring, prune the new leader to a strong bud just above the second horizontal wire and take back the short laterals you pruned in summer to the main stem. That summer, repeat the pruning steps for the first summer. Continue this regime in spring and summer until all the tiers are filled.

To maintain the shape thereafter, in summer cut sideshoots growing from the horizontal laterals to three leaves from the stem base. Remove completely any overly vigorous shoots growing from the horizontal laterals as well as any lateral shoots on the main stem.

Tie in ceanothus stems to form a fan shape against a trellis.

FAN-TRAINED SHRUBS

You can produce a simple and decorative fan-shaped feature by training the stems of some wall shrubs (see the A–Z, pp.40–141). To achieve this effect, follow the advice for training a climber on horizontal wires (see *opposite*), but in summer keep the stems tied to the fan of canes. When the stems outgrow the canes, attach them to the wires while maintaining the fan shape. You can also achieve this effect by tying the stems to a trellis panel.

TRAINING A BRAIDED STANDARD

Standard bay trees (*Laurus nobilis*) with plaited stems are expensive to buy but easy to make for a fraction of the cost. Select a young container-grown plant made up of lots of separate flexible stems – you will probably be able to make two standards from one purchase. Tip the plant out of its pot and use a sharp knife or spade to sever the rootball. Separate it into individual stems, each with several healthy roots attached. Plant three stems about 10cm (4in) apart in a large pot of soil-based potting compost or directly into the ground. Remove the leaves and shoots from the lower two-thirds of each stem so each is clear. Plait the clear stem segments together and tie them at the top with soft twine. Trim the growing tips on the stems on top to encourage bushy growth. As the plant develops, remove the twine and any growths from the plaited stems. Tip-prune the leafy stems on the top growth to maintain a rounded canopy.

Twist two or plait three bay stems together to create a braided standard.

When trimming box (*Buxus*) hedges or other shapes, it is better to use hand tools such as topiary shears rather than electric trimmers, to avoid bruising the leaves and to have finer control over the finished shape.

PRUNING AND TRAINING A–Z

Figuring out how and when to prune different plants can be baffling. There are general rules, but many climbers, shrubs, or trees require a more individual approach. Perhaps you have a garden full of overgrown specimens, or there is that one shrub you would like to flower more or would love to make more vigorous, but you don't know where to start. With our A–Z explaining exactly how to handle the most common garden plants, you can stop hesitating – and start pruning.

ABELIA *ABELIA*

Abelias are popular for their glossy rounded foliage and clusters of white, pink, or cerise, trumpet-shaped flowers, which bloom from summer to autumn, and are often scented. They are also appreciated for their usually red or pink calyces, which frame the blooms even when fading.

PLANT TYPE Deciduous and evergreen shrubs
HEIGHT 5m (16ft) or more
SPREAD Up to 4m (13ft)
FLOWERS ON Current year's stems, in summer–autumn
LEAF ARRANGEMENT Opposite or occasionally in whorls
WHEN TO PRUNE Deciduous in late winter–early spring; evergreens after flowering
RENOVATION Yes

HOW THEY GROW

Depending on the variety, abelias range in height but all have a fairly open structure and produce long, thin, arching stems. They tolerate a wide range of soil types, are generally trouble-free, and are ideal for a sunny border. Some are borderline frost hardy, down to -5°C (23°F), such as *A.* × *grandiflora* (syn. *A. rupestris* of gardens), or half-hardy, down to 0°C (32°F), like *A. floribunda* (syn. *Vesalea floribunda*), so need a sheltered position: for example, by being planted near a warm wall. Some types make an ideal informal hedge when trimmed regularly.

HOW TO PRUNE

There are three ways of pruning abelia on a regular basis, depending on the variety, along with a renovation pruning

Abelias are trouble-free shrubs with scented, funnel-shaped flowers with striking, darker-coloured bases.

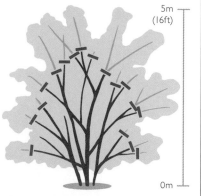

5m (16ft)

0m

In late winter or spring it is easy to spot the main framework on deciduous plants.

technique to rejuvenate the plant. Regular trimming will create a more dense plant, making it more suitable for a hedge. Deadhead regularly if it's practical to do so.

VIGOROUS DECIDUOUS TYPES In late winter or early spring, cut back hard to a permanent framework.

LESS VIGOROUS DECIDUOUS TYPES In late winter or early spring, remove misplaced shoots to maintain a good shape to the abelia.

EVERGREEN TYPES After flowering, lightly trim back shoots, so that the plant is well-balanced.

RENOVATION PRUNING In late winter or early spring, cut all flowered stems on overgrown plants back to a framework close to the ground. Use a pruning saw on large branches and loppers on smaller ones. After pruning, apply fertilizer and water well, then mulch. *Abelia* × *grandiflora* varieties benefit from removing one-third of older stems every three years.

ABUTILON *ABUTILON*

These plants have attractive, maple-like leaves and produce cheerful, generally bell- or cup-shaped flowers, typically in shades of red, yellow, and orange, but also white and blue. They flower for a long period, with the eye-catching, pendulous blooms hanging from their arching stems.

PLANT TYPE Deciduous and evergreen small trees, shrubs, perennials, and annuals
HEIGHT Up to 5m (16ft)
SPREAD Up to 2.5m (8ft)
FLOWERS ON New stems, in spring–autumn
LEAF ARRANGEMENT Alternate
WHEN TO PRUNE Winter–spring, or after flowering if wall-trained
RENOVATION No

HOW THEY GROW

The most commonly grown abutilons are half-hardy to frost-hardy shrubs, which can be planted outside in a sunny position. In areas with temperatures below 5°C (23°F), bring plants indoors for winter, into a frost-free place with plenty of winter light, such as a conservatory or unheated greenhouse. Evergreen to semi-evergreen A. 'Kentish Belle' has been known to keep producing its dangling bell flowers of red and apricot-yellow throughout winter when grown under cover. Abutilon × suntense 'Violetta' is usually deciduous and bears purple-blue flowers in early summer.

HOW TO TRAIN

Abutilons have rather spindly stems, which often require support with garden canes. If training against a wall, attach wires or a trellis to the wall and tie stems to these.

HOW TO PRUNE

Depending on the climate, abutilons are deciduous or semi-evergreen, or semi-evergreen or evergreen plants. A few types, like A. *megapotamicum* or A. *vitifolium*, need minimal pruning just to keep a good shape, in spring. Most others can be cut back harder (see

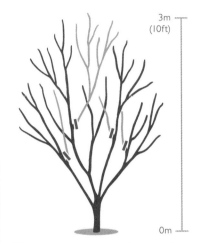

3m (10ft)

0m

Always remove crossing shoots back to their point of origin.

below). If in doubt, check the plant label for your variety. Stems of variegated species that revert – that is, produce shoots with plain green leaves – should be cut out at their point of origin.

DECIDUOUS TYPES Remove dead stems and crossing shoots in early spring, and, if required, prune back hard to a low established framework.

EVERGREEN TYPES In mid- or late spring, lightly trim shoots to shape. To reduce the size, cut branches back by no more than one-third, to above a node.

WALL-TRAINED TYPES After flowering or in late winter, prune flowered shoots back to 2–4 buds on a permanent framework of stems against the wall.

Abutilon **'Souvenir de Bonn'** has creamy white- to yellow-marked leaves.

Abutilon **'Kentish Belle'** produces apricot-yellow flowers with red above.

MAPLE *ACER*

Maples are much valued for their palmate (lobed) foliage, which can change to spectacular colours in autumn. Nearly all maples are deciduous, and the fresh young leaves unfurling in spring put on another show. Some varieties also provide winter interest, with coloured stems or patterned bark.

PLANT TYPE Deciduous and evergreen trees and shrubs
HEIGHT Up to 20m (65ft)
SPREAD Up to 12m (40ft)
FLOWERS ON New stems, in spring
LEAF ARRANGEMENT Alternate
WHEN TO PRUNE Autumn–winter
RENOVATION Yes

Acer crataegifolium **'Veitchii'** turns buttery-gold in autumn.

20m (65ft)

0m

Cut out dead or diseased wood and trim back new shoots to keep the tree to size.

HOW THEY GROW

The growth habit for most maples is fairly open, while some have a weeping or semi-weeping shape. They can develop into fairly large, upright trees, depending on the variety, but there are sculptural, mound-forming types as well. All maples prefer slightly acidic soil but will grow in any moist, well-drained soil in sun or partial shade. Position plants with delicate foliage, such as the lacy leaves of Dissectum varieties of

> **TOP TIP** RED- OR PURPLE-LEAVED MAPLES NEED SUN TO BOOST THE DEPTH OF FOLIAGE COLOUR, BUT PALER-LEAVED TYPES PREFER DAPPLED SHADE.

Japanese maple (*A. palmatum*), in a sheltered spot where they are protected from cold winds and late frosts, which can scorch. Tall maples may require staking in exposed positions. Smaller specimens can be grown in containers, but they need more attention than if planted in open ground.

FREE-GROWING TYPES Generally, maples require little pruning and are best left to grow naturally so that the beautiful shape of each individual plant can be appreciated. Some varieties have very low-growing branches, which touch the ground. In most cases this is fine, but you can raise the canopy if needed, removing the lowest branches.

This makes it easier to underplant if there isn't much space in the garden, and crown raising can also bring attention to features such as the peeling bark of *A. griseum*. Alternatively you can thin out some of the branches so that a view can be opened up through the plant, or so that you can more easily admire features such as the stripy stems of *A. pensylvanicum*, while still retaining the overall natural form of the plant.

HOW TO PRUNE

Immediately after planting, prune your tree or shrub to form a basic framework of three, four, or five branches, depending on the size of the plant. After this, it needs minimal pruning, which is done between late autumn and midwinter. Remove any misplaced or crossing branches and cut out any dead or diseased shoots, to maintain the good balanced framework. If a plant is getting too large for its space, cut back young shoots every year while maintaining a good shape to the plant.

RENOVATION PRUNING In exceptional cases, maples can be pruned very hard, back to 30cm (12in) above ground level; do this in autumn or winter. On small Japanese maples grown as specimens or weeping forms, look out for a knobbly graft union towards the bottom of the main stem and be careful not to cut below this. Never do any major pruning in spring (when the sap is beginning to rise) or in summer, because this can cause maples to "bleed" sap.

Thin out the branches on an *Acer negundo* back to the main stem.

Acer palmatum **'Ben-kagami'** foliage turns purple, maroon, and then scarlet.

MULTI-STEMMED TREE

Maples are generally grown with a central leading stem, which produces branches in the normal way, but you can also prune them to form a multi-stemmed tree. This adds to the attraction of the tree – especially those varieties that have special features such as A. *rubrum* 'Columnare' with its narrow upright habit.

Trim tips

Cut out central stem above opposite branches

Remove lowest branches

Prune out damaged growth

Cut back weak branches

Remove any stems showing dieback

Cut away branches that spoil the shape of the plant

Year 1 In winter, prune a young tree to around 50cm (20in), cutting back to two pairs of vigorous growths. Remove the lowest sideshoots.

Year 2 In winter, remove any weak and dead shoots, cutting just above healthy buds. Prune weak stems hard to encourage strong growth.

Year 3 In winter, remove lower shoots to expose attractive bark markings. Prune out any overcrowded shoots to give an overall balanced look.

SNOWY MESPILUS *AMELANCHIER*

This superb garden plant with spring to autumn interest is prized for its star- or saucer-shaped, pink to white flowers, which contrast beautifully with the emerging, bronze-pink young leaves. There is also spectacular autumn foliage, and small, maroon to purple-black fruits that are loved by birds. Though elegant, they are tough and hardy, tolerating a range of conditions including damp sites and shade.

PLANT TYPE Deciduous trees and shrubs
HEIGHT Up to 8m (25ft)
SPREAD Up to 8m (25ft)
FLOWERS ON Last year's stems, in spring–early summer
LEAF ARRANGEMENT Alternate
WHEN TO PRUNE Late winter
RENOVATION Partial

Amelanchier x grandiflora **'Robin Hill'** forms a tree up to 8m (25ft) tall.

HOW THEY GROW

Snowy mespilus makes an excellent specimen tree or border shrub. It is best planted in acidic soil in sun or partial shade, although *A. asiatica* is lime-tolerant. Most snowy mespilus develop a fairly light and airy growth habit, but some have a more spreading structure. One of the best for its open habit is *A.* x *grandiflora* 'Ballerina'. Its leaves are bronze-tinted when young, becoming mid-green in summer, and it bears white flowers on arching stems in mid-spring.

MULTI-STEMMED TYPES Snowy mespilus is particularly attractive when grown as a multi-stemmed tree. This is where three or more main stems develop from one point, giving a stunning effect. They can be trained this way as trees or be kept pruned when necessary as shrubs – a multi-functional plant.

HOW TO PRUNE

In late winter, remove any misplaced or crossing shoots to create a good balanced framework. Other than that, regular pruning isn't advisable, because these are fairly slow-growing trees, suitable for small and large gardens.

8m (25ft)

0m

To thin growth or create a multi-stem tree, cut stems back to near the ground.

Prune stems of *Amelanchier stolonifera* in need of renovation, using a pruning saw.

RENOVATION PRUNING A few species such as *A. stolonifera* can produce suckers or multiple stems around the base of the plant, becoming thicket-like. To remove these, prune hard in winter by shortening stems back to 30cm (12in) long. Alternatively, "stool" them by cutting all stems back to the ground in one go, or one-half one year and the other half the next. These plants are vigorous enough to rejuvenate quickly, and, although the new growth won't flower the season after renovation, the autumn leaf colour will be excellent.

ARTEMISIA _ARTEMISIA_

This group of plants is grown for its clouds of silvery aromatic leaves, as opposed to its insignificant yellow summer flowers, which can be trimmed off to encourage more foliage to develop. It is an excellent choice for adding texture to a Mediterranean-style planting scheme or dry garden.

PLANT TYPE Evergreen shrubs, perennials, and annuals
HEIGHT Up to 1.5m (5ft)
SPREAD Up to 1.5m (5ft)
FLOWERS ON Current year's stems, in summer
LEAF ARRANGEMENT Alternate
WHEN TO PRUNE Spring
RENOVATION Occasionally

HOW THEY GROW

Artemisias prefer a sunny site, generally with well-drained soil, but some species such as A. _lactiflora_ need fairly moist conditions. Most types will be short-lived if planted in heavy clay soils. Depending on the climate they grow in, some evergreen artemisias might act as semi-evergreens or deciduous plants, losing some or all of their foliage in winter, before producing fresh growth.

HOW TO PRUNE

To prevent plants becoming leggy over time, prune in spring after the threat of severe frost has passed. During the first spring after planting, shorten each stem to 2.5–5cm (1–2in) above soil level, to trigger strong new growth. Encourage a bushy growth habit by regularly pinching out the growing tips. On more mature plants, cut back the previous season's growth by one-half every year.

PERENNIALS Cut back perennial species such as A. _alba_ 'Canescens' to 2.5cm (1in) above ground level in spring.

BUSHY SHRUBS Shrub or subshrub species such as A. 'Powis Castle' can grow to around 1m (3ft) high and have a spread of a similar size. If they are getting too large, remove one-third of the previous season's growth in spring.

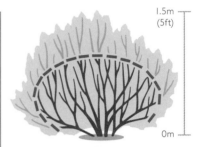

1.5m (5ft)

0m

Prune mature shrubby types back by one-third of the previous year's growth.

RENOVATION PRUNING If overgrown, leggy, and bare towards the base, you can attempt to rejuvenate a shrubby artemisia by cutting it back hard in spring, to within 5cm (2in) of ground level.

Southernwood (_Artemisia abrotanum_) produces fine, divided, grey-green foliage.

Artemisia 'Powis Castle' likes a position in full sun for its aromatic ferny foliage. It is only frost hardy so may not survive a severe winter.

SPOTTED LAUREL *AUCUBA JAPONICA*

These evergreen shrubs are popular mainly for their large glossy foliage. Some have variegated leaves, sometimes spotted, which add to the attraction, as do the red fruits produced in autumn. Being versatile and resilient, spotted laurel tolerates a wide variety of soil conditions and most situations within the garden, including shady areas. It also copes well with environmental challenges such as pollution.

PLANT TYPE Evergreen shrubs
HEIGHT Up to 3m (10ft)
SPREAD 3m (10ft)
FLOWERS ON Last year's stems, in mid-spring
LEAF ARRANGEMENT Alternate
WHEN TO PRUNE Late winter–spring
RENOVATION Yes

HOW THEY GROW

Spotted laurel tolerates salty coastal winds, full shade, and dry soil. It naturally develops a rounded bushy shape, which is good as a dense, semi-formal hedge. However, if it is trimmed regularly as a hedge, you will have to forego the red fruits in autumn. This shrub is prone to dieback and becoming bare at the base.

HOW TO PRUNE

After planting, shorten each stem by two-thirds, to encourage a bushy habit. Thereafter, as the plant matures, prune cultivars that bear berries in mid-spring, after the winter berries have faded or been eaten by birds. Remove any strong

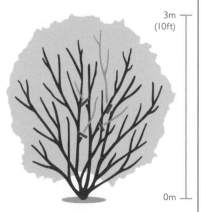

Trim back stems to healthy growth and cut out unwanted vertical branches.

Aucuba japonica **'Crotonifolia'** leaves look as if splashed with yellow paint.

vertical branches, to maintain a good shape. Cut out any shoots showing signs of dieback to healthy growth.

HEDGES If growing spotted laurel as a hedge, you can prune it in winter or spring by trimming with secateurs. Avoid a mechanical hedge trimmer, which may cut through the leaves, causing their edges to turn brown.

RENOVATION PRUNING Prune shrubs that are overgrown or have become bare at the base in several stages, over a period

Spotted laurel produces bright red berries in autumn when free-growing.

of two or three years. Never undertake such drastic pruning in a single stage as some shrubs can die.

In the first year, take out two or three main, or thicker, stems and trim some sideshoots to thin out the plants. Water, feed, and mulch with organic matter after pruning to encourage strong growth. New shoots will then develop from the pruned stems. In the second year, remove further mature stems and again trim any sideshoots. Water, feed, and mulch, as before. In the third year, repeat the pruning process once again.

BARBERRY *BERBERIS*

All barberries have spiny stems, making them ideal as an impenetrable barrier such as a boundary hedge, and many have eye-catching foliage in shades of red, as well as yellow or orange flowers. Varied in size and form, from large shrubs to small specimens, you can find one for almost any situation.

PLANT TYPE Deciduous and evergreen shrubs
HEIGHT Up to 4m (13ft)
SPREAD Up to 3m (10ft)
FLOWERS ON Old and new stems, in spring–summer
LEAF ARRANGEMENT In groups of three in axils of stem spines
WHEN TO PRUNE Deciduous after flowering and in winter; evergreens after flowering, generally
RENOVATION Yes

HOW THEY GROW

Barberries develop very dense growth and vary in shape from rounded to semi-prostrate, while some are upright or branching. They tolerate a range of soil types, provided it is humus-rich, in sun or partial shade. Barberries have yellow to dark orange flowers in spring and summer, and some of the deciduous species develop fine, autumn-coloured leaves, which are at their best if the plant is in full sun.

HOW TO PRUNE

The easiest way to prune barberries is with a hedge trimmer or shears. Thicker stems can be cut with loppers or a pruning saw. Always wear thornproof gloves and eye protection. Pruning is done at different times of year, depending on whether it is an

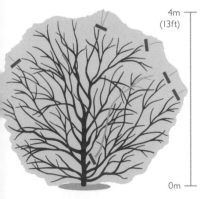

Shorten flowered stems and cut out bare wood on evergreen barberries.

Evergreen *Berberis darwinii* has orange flowers followed by blue-black fruits.

evergreen or deciduous plant, whether you want berries or not, and whether the plant is a hedge. You can trim hedges two or three times a year.

EVERGREEN TYPES Prune in spring or early summer, after flowering, provided you don't want berries. Cut out older stems and bare wood, and trim flowered shoots back to maintain the allotted size and encourage dense growth. If you want the plant to bear berries, prune in winter.

DECIDUOUS TYPES Prune from mid- to late winter for the best leaf show on specimen shrubs, cutting stems back to

near the ground. In summer, when you can more easily spot healthy growth, remove one in five stems, starting with dead, old, and bare-based ones.

RENOVATION PRUNING It can be difficult to thin out barberry, as it grows so densely and with spines on the branches. To renovate an overgrown barberry, cut it back hard to around 30cm (12in) above soil level, in late winter or early spring. Feed, water, and mulch with organic matter after renovating, to encourage strong growth. Note that you will lose flowers and berries for the first year after renovating your barberry shrub.

When grown in sun, *Berberis thunbergii* 'Orange Rocket' develops ruby foliage.

BIRCH *BETULA*

These deciduous trees and shrubs develop a graceful habit, lovely autumn foliage colour, and attractive male catkins in spring. Several types have striking white or peeling brown bark for added winter interest. Some are suitable for small gardens when grown singly or in groups of three or more.

PLANT TYPE Deciduous trees
HEIGHT Up to 20m (65ft)
SPREAD 10m (33ft)
FLOWERS ON New stems, in spring
LEAF ARRANGEMENT Alternate
WHEN TO PRUNE Late summer–winter
RENOVATION Yes

HOW THEY GROW

Birches are easy to grow in any fertile, well-drained soil in full sun or partial shade. They have a fairly open branch structure, which gives them an airy feel, and a light canopy that casts dappled rather than dense shade. While some, such as Erman's birch (*B. ermanii*), are wide-spreading in shape, with upward-reaching branches, others have a weeping habit. The downy birch (*B. pubescens*) prefers wetter conditions than the similar-looking silver birch (*B. pendula*), and does well in harsher climates at high northern latitudes.

Birches look wonderful in winter when planted in a group or as multi-stems to show off their trunks. Paper birch (*B. papyrifera*) is particularly appealing with its thin layers of peeling bark, which show pale orange-brown when first exposed, while Himalayan birch (*B. utilis* var. *jacquemontii*) develops the most striking white trunk once established. To get a multi-stemmed specimen quickly, plant three seedlings or small trees in the same planting hole and allow each to retain its central leader.

HOW TO PRUNE

Birches "bleed" heavily if cut in early spring (the sap seeps out from pruning cuts), so prune only from late summer to winter, when growth is slowing or the trees are dormant. Keep pruning to a minimum, removing only crossing branches and dead or diseased wood,

Silver birch can be grown as a multi-stem tree to show off its white bark.

while maintaining a balanced look to the tree. Generally, little other pruning is necessary unless you wish to remove suckers or sideshoots, or thin out stems in winter to expose more of the attractive bark. For example, river birch (*B. nigra*) has beautifully coloured, shaggily peeling bark, which is more noticeable when grown with a clear main stem.

WEEPING TYPES These can have their crowns lifted (*see p.31 for details*), which makes it easier to mow grass under each tree if grown as an individual specimen. Suitable plants to do this on include compact *B. pendula* 'Youngii',

with its arching branches, and the taller elegant 'Tristis', with its narrow frame and drooping foliage.

SIZE CONTROL You should reduce the crown of a birch tree for space only if absolutely necessary, and by no more than one-quarter of the branches at one time, because otherwise it can damage the tree. This work involves expertise, climbing up to cut back the tallest branches at the top of the tree, and is best done by a professional. If you are choosing a new birch tree to plant, consider a dwarf variety that would be more suitable for the space, so you won't have to prune it for size.

20m (65ft)

0m

Remove crossing branches as well as dead or damaged ones.

BRACHYGLOTTIS *BRACHYGLOTTIS*

Grown for their foliage and flowers, these evergreens are ideal for exposed and coastal gardens. *Brachyglottis* includes many plants formerly known as *Senecio*, and some herbaceous perennials and climbers, but most grown in the garden are mound-forming shrubs.

PLANT TYPE Evergreen small trees and shrubs
HEIGHT 1–5m (3–16ft)
SPREAD 2–4m (6½–13ft)
FLOWERS ON Previous and current year's growth, in summer–autumn
LEAF ARRANGEMENT Alternate
WHEN TO PRUNE After flowering
RENOVATION Yes

Brachyglottis Dunedin Group **'Sunshine'** bears long-lasting flowers.

The leaves of *Brachyglottis rotundifolia* are glossy green on top with buff undersides.

Deadhead and shorten stems after flowering to keep a compact shape.

HOW THEY GROW

The height and spread of different types of brachyglottis vary considerably. *Brachyglottis repanda* has a spreading habit and can grow to 3m (10ft) tall and wide, while more upright species such as *B. huntii* can be planted as a tree up to 5m (16ft) tall. Although usually vigorous, there are smaller species that are suitable for a rock garden or a small border. 'Sunshine', the most popular variety, belongs to a set of small bushy cultivars called the Dunedin Group, which grow up to 1.5m (5ft) tall and 2m (6½ft) wide, and can be used to create an interesting informal hedge.

The characterful leaves, with their split personality of glossy green on the upper side and hairy or buff silver-grey on the underside, provide a good show year-round. They are joined during summer by large, daisy-like flowers in yellow or white.

Brachyglottis is best planted in well-drained soil in full sun, though it tolerates partial shade. Being tough enough to handle the bracing windy conditions of coastal areas, this genus, especially *B. rotundifolia*, is a good choice for gardens by the sea, because it can withstand salt spray, and is drought tolerant. It will act as a good windbreak in exposed inland locations as well.

HOW TO PRUNE

Prune after flowering by lightly trimming or shortening stems that spoil the shape of the plant. Deadhead regularly where practical. Trim members of the Dunedin Group including 'Sunshine' at least once a year, to keep them compact, otherwise they might start to smother other plants. Alternatively you can take advantage of the spreading habit to keep down weeds, using them as ground cover.

RENOVATION PRUNING Cut woody stems right back to within 15–30cm (6–12in) of the ground. After such treatment it may not flower the following season.

'Sunshine' regenerates well with fresh growth after being cut back hard.

BUDDLEJA *BUDDLEJA*

These vigorous shrubs with an arching growth habit are renowned for their panicles of fragrant flowers. The most well-known species, *B. davidii*, is called the butterfly bush because of its magnetic appeal for pollinators, but can put on huge growth very quickly in one season if not kept in check.

PLANT TYPE Deciduous, semi-evergreen, and evergreen shrubs
HEIGHT Up to 9m (28ft)
SPREAD Up to 5m (16ft)
FLOWERS ON Current year's stems, in spring or summer, with some exceptions
LEAF ARRANGEMENT Opposite
WHEN TO PRUNE Spring
RENOVATION Yes

HOW THEY GROW

Buddlejas are mostly grown as freestanding shrubs or in mixed borders with other shrubs and herbaceous perennials. In late summer, butterfly bushes produce terminal flowers (at the end of the stem) that are long and conical, with contrasting coloured eyes, on the current year's growth. 'Pink Delight', with its sweet, honey-scented blooms in summer, is a compact cultivar that spreads up to 4m (13ft), while the hybrid 'Lochinch' grows 2–3m (7–10ft) high and wide, and has lavender-blue flowers with orange eyes in late summer. In contrast, *B. globosa* bears small, globe-shaped, yellow and orange flowers on the previous year's wood, in late spring or early summer.

Buddleja **'Lochinch'** is a good choice for a small garden as it remains compact.

The rounded flowers of *Buddleja globosa* are beautifully scented.

Plant buddlejas in a sunny spot with well-drained soil, as they hate being wet. They can self-seed prolifically and become invasive, so deadhead the inflorescences when they start to fade. This can result in a second flush of smaller flowers later in the season.

HOW TO TRAIN

Buddleja alternifolia has a lovely arching habit, so is often trained as a weeping standard, with the full length of its trailing stems packed with clusters of lilac flowers in early summer. Himalayan butterfly bush (*B. crispa*) is upright and likes some shelter so is ideal for wall-side borders and training against a wall.

HOW TO PRUNE

SPRING FLOWERERS Those semi-evergreen or deciduous types that flower in spring on the previous year's growth, including *B. globosa* and

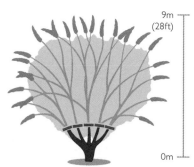

9m (28ft)

0m

Deciduous summer-flowering buddlejas can have all shoots cut back hard.

B. alternifolia, require a minimum of pruning. Trim to keep in shape only if necessary, after flowering in spring.

SUMMER FLOWERERS In early spring, cut all shoots to 30–45cm (12–18in) above soil level on vigorous deciduous types that flower in summer on the current season's growth, like *B. davidii* and its cultivars. A second prune to the same height in late spring has been proven to produce more flowers all over the plant, as well as keep it to a manageable size.

RENOVATION PRUNING Neglected summer-flowering plants can be pruned back hard in early spring, as above.

To restore spring-flowering types, after flowering remove old, weak, or dead stems and thin out by removing one-quarter of the stems to the base. Prune the flowered stems back to non-flowering shoots, preferably to upward- and outward-facing shoots. If not available, cut to a healthy bud.

BOX *BUXUS*

With its small evergreen leaves, box has historically been the top choice for parterres, knot gardens, and neat low hedging, providing structure in the garden throughout the year. This versatile shrub can also be clipped into balls, cones, and other topiary shapes grown in the ground or planted in containers.

PLANT TYPE Evergreen trees and shrubs
HEIGHT Up to 5m (10ft)
SPREAD Up to 5m 10ft)
FLOWERS ON Last year's and current year's stems, in spring
LEAF ARRANGEMENT Opposite
WHEN TO PRUNE Spring–summer
RENOVATION Yes

Common box is dense-growing, so perfect for clipping into balls and topiary.

HOW THEY GROW

Box tolerates most types of soil but suffers in wet and windy areas. Common box (*B. sempervirens*) is slow-growing, but will become tall and bushy without trimming to shape. The dwarf variety 'Suffruticosa' is perfect for creating a low hedge to edge a border or bed. Unfortunately *B. sempervirens* itself is vulnerable to box blight. As well as being careful when pruning, this risk can be mitigated by giving regular box feed, mulching in spring, watering at the base only, and clearing up fallen leaf litter from around the plant.

Buxus microphylla cultivars such as 'Faulkner', with smaller, glossy, deep green leaves, are more resistant to disease, and variegated types like *B. sempervirens* 'Elegantissima' and 'Variegata' also appear to be less affected. However, the box tree moth caterpillar is a serious pest that will infest and strip any type of box – always look out for it.

HOW TO PRUNE

Always cut box with hand tools. Using powered trimmers may give a tighter-looking result, but they can bruise the foliage, and the dense finish reduces airflow, which can encourage fungal diseases like blight. Trim plants on an overcast day if possible, so the sun doesn't scorch the cut leaves. Wipe off the sap and wash the tool blades as you work; sterilize them afterwards, to prevent the spread of fungal spores.

FORMAL SHAPES For a crisp finish, trim box little and often, with a first cut in late spring to early summer, and another in late summer to early autumn, with minor shaping trims as needed at other times: for example, when forming a

For best results, use clean sharp hand or hedge shears to trim and shape box.

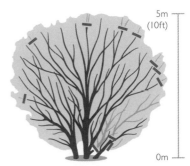

5m (10ft)

0m

Shorten new shoots to one or two leaves and cut out diseased stems.

new topiary. Intricate shapes can be cut with pruning secateurs or topiary shears, to give optimum control, while hedging and large cubes or flat-sided forms can be cut with hedging shears. To maintain the established shape, clip fresh green growth back to within two leaves of the darker old growth.

RENOVATION PRUNING Cut back hard to old wood, even to within 15–30cm (6–12in) of the ground, to help stimulate new growth or reshape a neglected structural form. For large plantings such as hedges, cut back one side the first year and let it recover, before cutting the other side the following year. Do renovation pruning in spring, once the risk of hard frost has passed.

BLIGHT SIGHT If leaves turn brown and drop in big sections – a typical sign of box blight – cut out all the affected area and burn the infected material; never put it on the compost heap.

BEAUTY BERRY *CALLICARPA*

These small trees and shrubs reveal their wow factor in autumn with eye-catching clusters of bright berries, which can last right through winter; in that way, the plant lives up to its common name. Some species produce attractive coloured foliage in spring and autumn.

PLANT TYPE Deciduous and evergreen shrubs and small trees
HEIGHT Up to 3m (10ft)
SPREAD 1–2m (3–6½ft)
FLOWERS ON Current year's stems, in summer
LEAF ARRANGEMENT Opposite
WHEN TO PRUNE Early spring
RENOVATION Yes

HOW THEY GROW

Beauty berries are ideal plants for a shrub border and grow well in fertile, well-drained soil that isn't too alkaline. They thrive in sun or partial shade, though they tend to grow slightly leggy in shady areas, and produce their best berries after a long hot summer. Plant together in groups to ensure cross-pollination, resulting in plenty of fruit.

A popular choice is *C. bodinieri* var. *giraldii* 'Profusion', which has bronze young leaves, purple or pink flowers, and metallic, dark-violet berries. *Callicarpa dichotoma* has long slender branches that arch down towards the ground, and bunches of violet-hued berries studded regularly along the stems. *Callicarpa japonica* 'Leucocarpa' bears white berries and golden autumn foliage. Most beauty berries are hardy, but some, like *C. rubella*, require frost-free conditions when growing outdoors. The berried branches can be cut and brought indoors for seasonal arrangements.

HOW TO PRUNE

Beauty berries can be left to establish in their natural form and reach their mature height and spread, with only

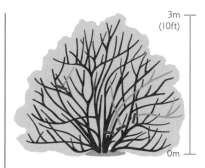

Remove inward-growing and crossing stems and those affected by winter dieback.

an occasional snip to keep the plants tidy and remove any overcrowded stems. Cut any crossing shoots that are rubbing against each other back to healthy buds, while trying not to spoil the symmetry of the plant.

Although not necessary, many gardeners choose to prune these shrubs every year to encourage a compact shape and more berries, cutting back hard to a framework of stems about 15cm (6in) from the ground. Time your pruning, in spring, between the last of the hard frosts and before fresh new growth appears.

RENOVATION PRUNING If old or growing lopsided, beauty berry bushes can be pruned hard as described above, and they respond well to such treatment. To rejuvenate an overgrown bush without losing its form, cut back one in five stems to the ground, starting with the oldest, every spring for a few years. Then mulch with well-rotted compost.

Violet-purple berries are borne on compact-growing *Callicarpa dichotoma*.

***Callicarpa bodinieri var. giraldii* 'Profusion'** flowers in midsummer.

BOTTLEBRUSH *CALLISTEMON*

Bottlebrush is so called because of its distinctive cylindrical flower spikes with long bristly stamens. They appear in spring and summer, in a range of colours, from bold shades of crimson, purple, and pink to white and gold, and are offset by aromatic leathery leaves.

PLANT TYPE Evergreen trees and shrubs
HEIGHT Up to 15m (49ft)
SPREAD Up to 5m (16ft)
FLOWERS ON Last year's and current year's stems, in spring–summer
LEAF ARRANGEMENT Alternate
WHEN TO PRUNE After flowering
RENOVATION Yes

HOW THEY GROW

These evergreens have a fairly dense, spreading growth habit, and prefer slightly acidic, moist soil. Although generally trouble-free, they prefer warm conditions in full sun, in a position free from severe frosts and cold winds. They do well in the ground in mild and coastal areas, but elsewhere are best planted in a sheltered border or at the base of a south- or west-facing wall or fence.

Green bottlebrush (*C. viridiflorus*) is hardier than most species, and can withstand temperatures down to -5°C (23°F). Half-hardy species such as the crimson bottlebrush (*C. citrinus*) and its scarlet-flowered varieties including 'Splendens' are best grown in containers so they can be brought indoors into frost-free conditions over winter. Bottlebrushes are somewhat drought-tolerant, and are susceptible to root rot, so be careful not to overwater.

HOW TO PRUNE

Prune lightly to retain an attractive balanced shape after flowering has finished, cutting back to just behind the spent flowerheads. The occasional shoot may grow out and into other

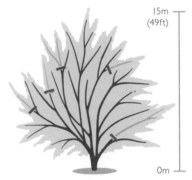

15m (49ft)

0m

Trim back shoots after flowering to maintain the shape of the plant.

plants, but these are easily dealt with by pruning back to a shoot or buds on a healthy stem, to retain the symmetry. You can also remove dead wood and stems that are crossing.

DEADHEADING Hard, grey-brown seed capsules develop in rows along the branches of bottlebrush, with a new set added each year. If you don't like how they look on the plant, simply deadhead the flowers regularly to avoid them setting seed. This will encourage more flowers too.

RENOVATION PRUNING Most bottlebrush plants tolerate hard pruning, so if any have outgrown their space cut them back in stages over several years. After flowering, remove two or three of the older stems to the base; and, to reduce the height, cut back the longest stems by one-third to younger, outward-facing shoots.

Callistemon rigidus (syn. *Melaleuca rigidus*) may need some support if left to grow naturally, and unpruned, in warmer climates.

HEATHER *CALLUNA*

Often seen growing in vast swathes of pink, red, white, and purple in the wild, heather also makes a wonderful garden plant, and its foliage generally changes colour over winter. Bees and other pollinators adore the flowers that smother these low-growing shrubs.

PLANT TYPE Evergreen shrubs
HEIGHT Up to 60cm (24in)
SPREAD 60cm (24in)
FLOWERS ON Current year's stems, in midsummer–late autumn
LEAF ARRANGEMENT Opposite
WHEN TO PRUNE Spring, before growth starts
RENOVATION No

HOW THEY GROW

Known as common heather, Scots heather, or ling, *C. vulgaris* thrives in an open sunny position, though it can manage some shade. For best results grow it in light acidic soil with good drainage. As it is fully hardy, it tolerates exposed sites. Its growth habit varies with cultivar, from upright to prostrate, but most heather will form a thick mat of foliage, which makes it ideal as ground cover to suppress weeds.

There are hundreds of cultivars to choose from, including 'Annemarie', which has a neat habit, long spikes of pink double flowers that pack its green stems from summer to autumn, and bronze winter foliage. 'Firefly' has striking terracotta foliage in summer, which turns brick-red over winter.

HOW TO PRUNE

Heather has a bushy dense habit, but can become straggly if not trimmed regularly – once a year is usually sufficient. During spring, before fresh growth starts, cut back the flowered stems to about 2.5cm (1in) above old growth, using a pair of hedge shears. Deadheading regularly during the flowering season also helps to keep the plant in shape.

Heathers may only last about ten years in a garden setting before becoming leggy, with all the growth and flowers at the top of the plant. However, they do not regrow reliably from old wood, so it is best not to cut back hard.

DROPPING If an old heather has become woody, instead of renovating through pruning, you can propagate it by "dropping" in spring. Dig it up and make a new planting hole large and deep enough to bury two-thirds of the whole plant. Pop it in the hole and backfill with a 50:50 mix of grit and coir, leaving the shoots sticking out at the top. Keep watered in dry weather in summer. Lift in autumn and identify rooted shoots; cut these off and plant as replacements for the discarded parent plant.

Common heather will grow in exposed areas as well as in more sheltered spots.

Regularly removing the dead flowerheads on heather will keep growth neat, as will trimming the plant again in spring.

CAMELLIA *CAMELLIA*

These woodland-edge shrubs with their glossy green leaves are prized for their elegant flowers during early spring, when most plants are still dormant. The blooms are produced in shades of pink, red, and white, appearing singly or in clusters, and they last for several weeks.

PLANT TYPE Evergreen shrubs
HEIGHT 5m (16ft)
SPREAD Up to 3m (10ft)
FLOWERS ON New stems, in spring
LEAF ARRANGEMENT Alternate
WHEN TO PRUNE After flowering
RENOVATION Yes

Camellia × williamsii **'Bowen Bryant'** is a large, upright, evergreen shrub.

Camellia japonica **'Nobilissima'** has white, peony-form, double flowers in spring.

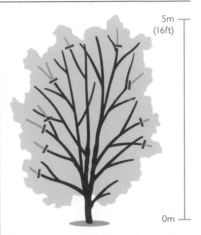

5m (16ft)

0m

Trim to required size and shape by cutting back overlong stems.

HOW THEY GROW

As woodland plants, camellias do best in partial or dappled shade, and are suitable for north-facing gardens. They prefer well-drained, neutral to acidic soil, but if your soil is alkaline you can still grow smaller varieties in containers filled with ericaceous (lime-free) compost.

There are several types of camellia flower, from single to rose and peony forms, and miniature to extremely large. Some are also scented. The most commonly grown types, such as *C. japonica* cultivars like 'Nobilissima' and *Camellia × williamsii* hybrids like 'Debbie', flower in early spring and are fully hardy, but some, for example *C. reticulata*, are half hardy and need a sheltered site in cool-temperate areas, or can be grown in pots brought under glass in winter.

However, even on the hardy camellias, buds and flowers are prone to browning from frost damage. To avoid this, plant in a site sheltered from cold winds and early morning sun. *Camellia sasanqua* is an exception in many ways: it thrives in a sunny spot, as long as its roots stay cool; and it flowers in autumn.

HOW TO PRUNE

It is not usually necessary to prune camellias – they have a pleasant form and don't require encouragement to flower. However, if you wish to keep your plant to a manageable size with a dense growth habit, you can trim it lightly each year with sharp secateurs. After blooming in spring, deadhead spent flowerheads and take out crossing or dead branches. Clip leggy stems back

to the main mass of the bush, and thin out crowded stems, to improve airflow. If there is an allotted size to which you wish to keep the plant, cut back stems to within 5cm (2in) of this framework.

BAD SPORT Some cultivars may produce a "sport", that is, a shoot with a different-coloured flower. This is a genetic throwback, so cut away these shoots to prevent the plant reverting to its original form.

RENOVATION PRUNING Camellias respond well to hard pruning, but you might lose one or two years' flowers as a result. To rejuvenate an overgrown shrub, cut all stems back to 45–60cm (18–24in) long, in spring once there is little risk of frost. New shoots will soon sprout from the old wood, and stems from the base.

BLUEBEARD _CARYOPTERIS_

Both flowers and foliage on these dainty shrubs exude a beautiful aromatic scent, like lavender. Their small fluffy flowers, produced in shades of blue from late summer through autumn, and in milder areas sometimes into winter, are loved by bees and other pollinators.

PLANT TYPE Deciduous shrubs and perennials
HEIGHT Up to 1.2m (4ft)
SPREAD 1.5m (5ft)
FLOWERS ON Current year's stems, in late summer–autumn
LEAF ARRANGEMENT Opposite
WHEN TO PRUNE Late winter–early spring
RENOVATION Yes

Bluebeard has grey-green foliage and produces bright blue flowers from late summer onwards, for several months.

When renovating, shorten stems to 15–30cm (6–12in) above ground level.

HOW THEY GROW

Bluebeard has an upright habit, and is drought-tolerant and generally trouble-free. Grow it in well-drained but fertile soil in full sun. It benefits from being planted against a warm wall, especially in areas where winter temperatures drop below -15°C (5°F) and summers are also generally cool.

The most well-known types, cultivars of _C. × clandonensis_ such as 'Heavenly Blue' and 'Kew Blue', are the result of an accidental cross. The distinctive flowers grow in clusters set around the long stems at intervals. The majority are blue, ranging in hue from purple and sapphire to deep blue, but some, like STEPHI, bear pink or white blooms.

Most bluebeards have toothed, greyish green foliage, though 'Worcester Gold' develops warm yellow tones, and 'White Surprise' has white-edged, green leaves.

HOW TO PRUNE

These shrubs can become leggy and bare at the base as they age, so prune regularly to keep a compact shape and to promote plenty of leaves and flowers. From late winter to early spring, when severe frosts are no longer a threat, many gardeners also clip back the previous year's flowering growth to within 2.5cm (1in) of the old wood framework, and then apply a mulch of well-rotted garden compost.

RENOVATION PRUNING If plants do become bare at the bases, prune all flowered stems to within 15–30cm (6–12in) of soil level, in late winter or early spring. This will result in good, compact, dense growth from the base.

Clip back stems to within 2.5cm (1in) of the permanent woody framework.

CALIFORNIA LILAC *CEANOTHUS*

California lilac flowers profusely with small dense clusters of intense blue, or sometimes white or pink, flowers produced at the tips of shoots or on small sideshoots from spring to autumn. It is a great shrub for coastal gardens, and also offers many benefits for wildlife.

PLANT TYPE Deciduous and evergreen shrubs
HEIGHT 4m (13ft) or more
SPREAD Up to 3m (10ft)
FLOWERS ON Last year's and current year's stems, in spring–autumn
LEAF ARRANGEMENT Alternate or opposite
WHEN TO PRUNE Spring–midsummer, depending on variety
RENOVATION No

HOW THEY GROW

The growth habit is generally dense and bushy, but some plants have a more prostrate habit, making them ideal for covering banks or unsightly objects like manhole covers; they also provide good, weed-suppressing ground cover.

Deciduous types of California lilac such as *C. × delileanus* 'Henri Desfossé' are hardier than the more common evergreen varieties, but all of them do best when planted in a sunny sheltered spot out of cold winds, such as against a south-facing wall or fence. They prefer well-drained, neutral soil and will suffer in wet conditions.

Evergreen *C.* 'Puget Blue' is sought after for its masses of flowers, which last for weeks in late spring and early summer; while hardy *C.* 'Concha', also evergreen, wears its brilliant blue clusters on cascading stems.

HOW TO TRAIN

Some cultivars such as *C.* 'Cynthia Postan' are suitable for training against a wall, usually in a fan shape. When grown against a warm wall, California lilac can grow taller and wider than as a freestanding shrub, so may need more regular pruning to keep it within bounds. Tie in regularly, and once established on a trellis or wire framework, prune at the appropriate time for the type (*see below*) by removing any wayward and dead stems, and clipping back sideshoots to within 2–4 buds of the main stems.

HOW TO PRUNE

EVERGREEN TYPES California lilacs require only minimal pruning but the timing of this depends on the variety. Prune those that flower in spring, such

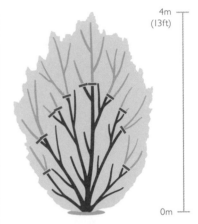

Prune one-third off flowered stems to keep a bushy shape.

as *C. arboreus* 'Trewithen Blue', after flowering, in midsummer. Prune those that bloom late in the year, such as *C.* 'Autumnal Blue', during early or mid-spring. In all varieties, remove dead, damaged, or diseased stems and, to keep a bushy shape, cut back the flowered stems by one-third.

DECIDUOUS TYPES Prune those that produce flowers on new growth from midsummer to autumn, such as *C. × delileanus* 'Gloire de Versailles', in early or mid-spring in the same way as evergreen types, to maintain a bushy shape and encourage more flowers.

PROSTRATE TYPES These (for example, *C. thyrsiflorus* var. *repens*) are best left unpruned, except where they are becoming a nuisance.

California lilac thrives in a sheltered site such as against a south-facing wall.

Clip sideshoots back to a few leaves from a main stem on a wall-trained plant.

CERATOSTIGMA *CERATOSTIGMA*

Ceratostigmas are beloved for their true blue flowers in shades from pale Wedgwood blue to deep indigo, with contrasting red bases. The star-like blooms appear from late summer to autumn, and are set off beautifully by the foliage turning from green to russet-red as the season changes.

PLANT TYPE Deciduous and evergreen shrubs and subshrubs
HEIGHT Up to 1m (3ft)
SPREAD 1.5m (5ft)
FLOWERS ON Current year's stems, in late summer–autumn
LEAF ARRANGEMENT Alternate
WHEN TO PRUNE Early spring–mid-spring
RENOVATION No

Ceratostigma plumbaginoides does well at the front of a border or on a slope.

HOW THEY GROW

Ceratostigmas are best grown in a sheltered sunny spot in light to medium, moist, fertile, well-drained soil, though they appreciate some partial shade during hot summers.

Deciduous *Ceratostigma willmottianum* develops a bushy spreading habit, which reaches about 1m (3ft) high. This shrub is decorated from late summer into autumn with terminal clusters of pale blue blooms with dark red bases.

Ceratostigma plumbaginoides is sometimes known as blue-flowered leadwort. It produces red wiry stems and vivid blue flowers with red bases,

and bright green leaves that tinge crimson in autumn. Depending on the climate where it grows, treat it as a herbaceous perennial or a woody-stemmed perennial or subshrub, based on whether its stems die back in the cold of winter. Being a lower-growing, creeping species often used for ground cover, *C. plumbaginoides* is suitable for growing at the front of a border or in a rock garden. It spreads through rhizomes, so is also helpful for stabilizing the soil on a slope or bank.

Less hardy species such as the semi-evergreen *C. griffithii* may need some protection such as a cloche over winter; otherwise grow it in a container and shelter it during severe weather.

HOW TO PRUNE

Prune in early or mid-spring, by cutting out damaged growth and trimming the flowered stems back to within 2.5cm (1in)

Ceratostigma willmottianum has blue star-shaped flowers surrounded by red.

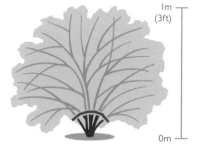

1m (3ft)

0m

Cut flowered stems back to a short framework of the older stems.

of the old growth. If *C. plumbaginoides* retains its stems over winter, trim these back in the same manner to a short framework; if the stems die down in autumn or winter, cut clean back to within 2.5–5cm (1–2in) of the ground.

Deadhead all plants after flowering to better show off the colourful autumn foliage – but as this plant can cause skin irritation always wear gloves when handling green material in this way.

The foliage of many species develops attractive shades of purple-red in autumn.

JAPANESE QUINCE *CHAENOMELES*

These versatile hardy shrubs are among the first to flower every year, and can be grown as a freestanding specimen or a hedge, or be trained against a wall or fence. The blossom is borne in small clusters along the bare branches before the leaves appear, and it is followed by apple-shaped fruits.

PLANT TYPE Deciduous shrubs
HEIGHT Up to 2.5m (8ft)
SPREAD Up to 5m (16ft)
FLOWERS ON Last year's stems, in late winter–early spring
LEAF ARRANGEMENT Alternate
WHEN TO PRUNE After flowering, in late spring; wall-trained also in summer
RENOVATION Yes

Chaenomeles × superba 'Crimson and Gold' has a compact habit.

HOW THEY GROW

Japanese quince has a spreading habit, and grows in full sun or partial shade in any reasonably fertile, well-drained soil. It even tolerates heavy clay, but suffers in alkaline soil, which may turn the foliage yellow (chlorotic).

Depending on the variety, the blossom can be single or double, and ranges in colour from deep red (like *Chaenomeles × superba* 'Crimson and Gold') to pink (*C. speciosa* 'Moerloosei') and white (*C. speciosa* 'Nivalis'). It is a welcome nectar source for wildlife early in the year, and makes a wonderful show before the toothed, dark green leaves appear. The fragrant autumn fruits are yellow to green, and are edible if cooked.

Japanese quince is trouble-free generally, but care should be taken

Flushed pink flowers are produced on *Chaenomeles speciosa* 'Moerloosei'.

when handling as many have spines on the branches. When cut, these plants sometimes send up suckers from the base, which can be pruned out or used for propagation.

HOW TO TRAIN

Quinces make excellent wall-trained plants on a sheltered sunny wall or fence. For a fan shape, attach wires horizontally to the wall, fixing them 20cm (8in) apart, and tie the stems to the wires, spacing them out like the spokes on a bicycle wheel.

HOW TO PRUNE

FREESTANDING TYPES When grown as a freestanding specimen, Japanese quince does not require much attention and can be left alone, but regular light pruning will ensure a neat appearance and improve flowering. Wear gauntlets and use anvil loppers to avoid injury from the vicious spines. After flowering, in late spring, remove dead, diseased, or damaged stems and any crowded or crossing branches completely, and cut back new growth to 4–6 leaves.

WALL-TRAINED TYPES After flowering, cut sideshoots back to 2–4 buds from the permanent framework of branches. You may need to do some summer pruning as well, cutting vigorous stems to about six buds from the permanent framework. Cut any shoots growing towards or away from the wall or fence, back to a main stem.

RENOVATION PRUNING If a Japanese quince has grown too large or leggy, it can be rejuvenated by pruning out one-third of the oldest stems right down to the ground after flowering, every year for three years.

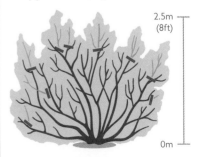

2.5m (8ft)

0m

Keep a neat shape by shortening the new growths after flowering.

MEXICAN ORANGE BLOSSOM *CHOISYA*

The glossy aromatic foliage of these evergreen shrubs offers year-round interest, and in addition they give a superb display with their clusters of white fragrant flowers in late spring and summer. They are compact and easy to grow.

PLANT TYPE Evergreen shrubs
HEIGHT 2.5m (8ft)
SPREAD 2.5m (8ft)
FLOWERS ON Last year's and current year's stems, in late spring–summer
LEAF ARRANGEMENT Opposite
WHEN TO PRUNE After flowering and in early spring
RENOVATION Yes

Choisya **'Aztec Pearl'** is a compact-growing shrub, which sometimes flowers twice a year.

HOW THEY GROW

Mexican orange blossom fits well into most gardens because it is a medium-sized, mound-forming plant that usually reaches about 1.2m (4ft), but can spread up to 2.5m (8ft) in ideal conditions. The growth habit is very dense, which makes it excellent as an informal hedge with just the occasional trim to keep it within bounds. It grows best in full sun in any well-drained soil, but tolerates dappled shade though the foliage won't be as colourful.

Choisya × *dewitteana* WHITE DAZZLER is hardier than most, while *C. dumosa* var. *arizonica* benefits from a sheltered position, with the protection of a wall, for example. Popular types include *C. ternata* and *C.* × *dewitteana* 'Aztec Pearl', which sometimes produce their star-shaped blooms twice in a season. *Choisya ternata* SUNDANCE rarely flowers, but this is more than made up for by its bright yellow to lime-green foliage.

HOW TO PRUNE

After flowering, cut back stems that spoil the shape of the shrub, and prune out damaged, dead, and diseased wood, stems that rub, and older stems that are bare at the base. Trim back the faded flower stems to encourage more flowers later in the year. You may also need to lightly trim any stems damaged by frost early in spring.

Over time, Mexican orange blossoms can grow leggy and unruly, with branches that break easily. Avoid this with regular trimming to keep the plant to whatever size is required – if not every year, then prune at least every three or four years. This will encourage a better shape, more flowers, and more intensely coloured leaves.

RENOVATION PRUNING If getting too large for its allotted space, Mexican orange blossom can be pruned hard. In early spring, cut out broken, diseased, dead, and crossing stems, and prune back the rest by at least one-half. For a more drastic renovation, shorten all stems back to within 30cm (12in) of the base of the plant. It may take a season or two to recover fully.

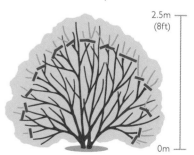

2.5m (8ft)

0m

After flowering, shorten flowered stems to trigger a second flush of blooms.

CLEMATIS *CLEMATIS*

These climbers are valued highly in the garden for their blooms. There are hundreds to choose from, varying from vigorous evergreen types that flower in early spring to deciduous, large-flowered cultivars that bloom into autumn. Use to clothe a fence or wall or to scramble up a tree.

PLANT TYPE Deciduous and evergreen climbers

HEIGHT Up to 14m (46ft)

SPREAD Up to 3m (10ft)

FLOWERS ON Last year's and current year's stems, in early spring–autumn

LEAF ARRANGEMENT Opposite or occasionally alternate

WHEN TO PRUNE Group 1 after flowering; groups 2 and 3 in early spring

RENOVATION Occasionally

HOW THEY GROW

Most varieties of clematis use their twining leaf stalks to attach themselves to their supports, or other plants. The roots prefer to be kept cool, so plant the base in shade in any fertile, well-drained soil in full sun or partial shade, and make sure the top of the rootball is about 8cm (3in) below the soil surface, to encourage the production of strong shoots from below soil level.

Clematis flowers vary widely in size, shape, and hue, from large flat blooms to small nodding bells, and in soft and

TOP TIP TO REDUCE THE CHANCE OF YOUR PLANT BEING AFFECTED BY THE FUNGAL DISEASE CLEMATIS WILT, GROW SHALLOW-ROOTED, LOW-GROWING PLANTS NEARBY TO SHADE THE BASE, OR SURROUND THE CLEMATIS WITH A MULCH OF PEBBLES TO KEEP IT COOL.

CLEMATIS GROUPS

There is such a broad range of different types of clematis that, as an aid for pruning, this genus has been divided into three groups based on when each plant flowers.

Group 1 These are early-flowering species and cultivars that bloom in early spring on the previous year's shoots. This group includes *C. montana*, a vigorous deciduous type, and *C. alpina*, which is excellent for cold and exposed situations, as well as *C. macropetala* and the evergreen 'Apple Blossom', which grows to 8m (26ft) and produces abundant, pinky peach-tinted flowers.

Group 2 This group contains deciduous cultivars that flower in late spring and early summer, on sideshoots produced in the previous year. They have large, often double or semi-double blooms. Some can be trained up an obelisk in a mixed border or be grown in a large container. The group includes cultivars like purple-flowered *C.* 'The President', pink-mauve *C.* 'Nelly Moser', and dark red *C.* 'Niobe'.

Group 3 All in this group bear flowers on the current year's growth. The larger-flowered hybrid types are deciduous, with single, saucer-shaped flowers borne in summer and early autumn, while the species and small-flowered hybrids flower from summer to late autumn. Popular types include *C. tangutica*, *C.* 'Polish Spirit', and *C.* 'Étoile Violette'.

Evergreen *Clematis armandii* bears scented white flowers in early spring.

Deciduous *Clematis* 'Niobe' flowers over a long period in summer.

Deciduous *Clematis* 'Etoile Violette' has profuse flowers from midsummer.

bold shades from white, gold, and orange to blues, pinks, and scarlets. Some, such as yellow-bloomed *C. tangutica*, also produce attractively silky seedheads.

HOW TO TRAIN

When planting a new clematis, make sure to offer support from the very beginning, providing a strong cane to guide stems to a trellis, arch, or tree. For all groups, encourage a good framework of stems by shortening all stems back to within 30cm (12in) of the ground in the first spring, to just above a pair of leaf buds.

As they grow, guide stems towards the main support structure, spacing stems out evenly to get broad coverage over the support and tying each in gently with twine or other soft ties. The twining leaves can only wrap themselves around objects 0.5cm (¼in) or less in diameter, so if your support system is thicker than this you may need to provide a wire grid for them or else manually tie in lots more stems.

HOW TO PRUNE

GROUP 1 Prune after flowering to remove dead or damaged shoots, and shorten stems to the allotted space. This encourages new growth in summer, which will bear flowers the following spring. Once established, vigorous types like *C. montana* need minimal pruning – just enough to stop them outgrowing their space.

GROUP 2 Once a permanent framework is established, prune in late winter or early spring, trimming weak shoots back to their point of growth, or cutting out entirely if they are damaged. Clip sideshoots back to healthy buds.

GROUP 3 Prune this group in late winter or early spring by cutting all the stems back to around 30cm (12in) above the

> **TOP TIP** WHEN PRUNING BACK A LARGE PLANT, IT MAY PROVE EASIER TO REMOVE MOST OF THE GROWTH WITH A PAIR OF SHEARS, AND THEN DO MORE DETAILED PRUNING WITH SECATEURS.

Prune a late-flowering Group 3 clematis back to near the base, cutting above a pair of fat, strong-growing buds.

ground, just above a pair of healthy buds. Make sure any dead wood is completely removed.

RENOVATION PRUNING Group 1 types of clematis, except for *C. armandii*, usually respond well to renovation, so if you are faced with an overgrown specimen you can cut it back hard almost to the base. Flowers will be lost for the first year after such hard pruning. Leave the plant for about three years before pruning again.

After planting, help guide young stems towards the tree, shrub, trellis, or other support, using a strong cane and soft ties.

Begin to prune a large clematis plant by cutting back the mass of top-growth with shears.

HOW TO PRUNE CLEMATIS BY GROUP

Vigorous Group 1 *Clematis montana* 'Marjorie'

Group 1 Cut out congested stems or growth that is exceeding the allotted space, by trimming each stem back to just above a healthy pair of buds. Cut any weak and damaged stems to strong buds lower down the stem.

Large-flowered Group 2 *Clematis* 'Vyvyan Pennell'

Group 2 Remove any weak and damaged growth from the point where it grows or back to ground level. Leave a permanent framework with evenly spaced shoots of strong stems. Shorten all sideshoots to healthy buds – these will produce flowering shoots.

Group 3 In early spring, remove any dead growth where buds are not developing. Prune remaining stems back to around 30cm (12in) from ground level.

Small-flowered Group 3 *Clematis* 'Bill MacKenzie'

CLERODENDRUM *CLERODENDRUM*

These vigorous shrubs start to produce their eye-catching, fragrant, white or pink flowers in summer, followed by blue berries with a striking maroon collar. Most are tender types from tropical and subtropical climates, but there are some hardier varieties that can be grown in cooler gardens.

PLANT TYPE Deciduous and evergreen small trees and shrubs
HEIGHT Up to 6m (20ft)
SPREAD Up to 6m (20ft)
FLOWERS ON Last year's and current year's stems, in late summer–autumn
LEAF ARRANGEMENT Opposite or occasionally in whorls
WHEN TO PRUNE Early spring
RENOVATION Yes

The blooms of glory flower are fragrant, while its leaves are pungent.

HOW THEY GROW

Clerodendrums are unusual shrubs with an upright bushy growth habit, and are often grown as feature specimens. Plant in rich, moist but well-drained soil in full sun in a sheltered position, by a warm wall perhaps, as they suffer dieback and leaf damage in exposed situations. They can also cope with some shade. Some types sucker freely.

The most popular varieties include harlequin glorybower (*C. trichotomum*), which has pink buds from which the white flowers, beloved by pollinators, develop in late summer. In autumn, the star-shaped calyces framing the petals turn dark pink or red, and hug the base of the turquoise berries. Harlequin glorybower is hardy to -10°C (14°F).

Harlequin glorybower has metallic blue berries with reddish pink collars.

Glory flower (*C. bungei*) is a more compact choice of clerodendrum, growing to about 2.5m (8ft), and is frost hardy to -5°C (23°F). It bears sweetly scented, soft pink, dome-like panicles of densely packed, star-shaped flowers in late summer. Its toothed, heart-shaped leaves have a strong sharp – some say unpleasant – smell when brushed against or crushed.

HOW TO PRUNE

Once established, clerodendrums need only minimal pruning. Keep a neat shape by trimming back wayward shoots and removing damaged growth in early spring. However, when left to their own devices, they can send up suckers from the base, forming a thicket. These growths also tend to pop up elsewhere in the garden, and should be pulled out. Carefully remove the soil around the base to expose where the sucker meets the plant, and tear it off – if you cut, it will produce yet more suckers.

To promote larger leaves, prune back strong-growing glory flower to a low permanent framework, about 60cm (24in) from the base, in early spring as growth begins. Some gardeners like to follow this with a second prune in midsummer, to trigger another flush of leaves, but this is likely to delay flowering until autumn.

RENOVATION PRUNING Cut back overgrown shrubs hard to fit their allotted space. In early spring, shorten all stems back to about 30cm (12in) from the ground, each time cutting just above a healthy bud.

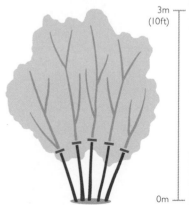

3m (10ft)

0m

To encourage large leaves, cut stems about 60cm (24in) above ground level.

DOGWOOD *CORNUS*

Decorative stems, beautiful flowers, and interesting leaves and growth habit are just some of the reasons that dogwoods are such highly valued garden plants. Some types also produce pretty fruits, while others have eye-catching autumn foliage colouring.

PLANT TYPE Deciduous trees and shrubs
HEIGHT Up to 7m (23ft)
SPREAD Up to 5m (16ft)
FLOWERS ON Last year's and current year's stems, in late spring–summer
LEAF ARRANGEMENT Opposite or occasionally alternate
WHEN TO PRUNE Late winter–mid-spring
RENOVATION Occasionally

HOW THEY GROW

Plant dogwoods in sun or partial shade. Some such as *C. kousa*, *C. nuttallii*, and their hybrids bear decorative "flowers" in summer, and are grown as specimen trees. Others such as wedding cake tree (*C. controversa* 'Variegata') have an airy tiered shape and captivating foliage. They prefer rich, well-drained soil.

Many shrubby cultivars are grown for their colourful stems in winter. *Cornus sanguinea* 'Midwinter Fire' and *C.s.* 'Anny's Winter Orange' flame orange and red in the low winter sun. *Cornus alba* 'Kesselringii' produces dark purple stems, while *C. sericea* 'Flaviramea' develops bright lime-yellow ones. These types tolerate a broad range of soil conditions, even damp areas, but colour up best in full sun.

The bright yellow stems of *Cornus sericea* 'Flaviramea' show up well in winter.

HOW TO PRUNE

The pruning requirements of dogwoods vary, depending on the reason they are being grown.

FOR FLOWERS OR FOLIAGE Minimal pruning is necessary for species like *C. alternifolia*, grown for its tiered shape and leaf colour, or *C. florida*, grown for its bracts. When dormant in late winter and early spring, trim growth only as needed to maintain symmetry, and to cut out dead wood. These types will not tolerate hard pruning.

FOR STEMS Prune varieties grown for their vividly coloured winter stems, such as *C. alba* 'Sibirica', in spring every year, or every two to three years for those in shaded areas. Previously it was

Coppice dogwoods with colourful stems like *Cornus sanguinea* each spring.

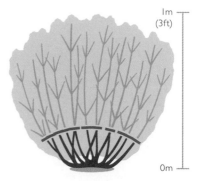

1m (3ft)

0m

If grown for stems, cut back to two buds above the previous year's growth.

always advised to prune them in early spring, but recent research indicates mid-spring is just as good a time, meaning that you can enjoy the stem colouring for longer in the year.

For a few years, refrain from pruning newly planted dogwoods grown for their stems. Then, once established, shorten new growth back to 8–10cm (3–4in) from the ground. In subsequent years, cut back to within two or three buds of the previous year's growth. Use sharp secateurs on thin stems and a pruning saw on thicker ones.

Such constant coppicing encourages new growth on the plant – the best-coloured stems – and means that the plant's shape will stay fairly compact.

RENOVATION PRUNING *Cornus officinalis* and *C. mas* plants that have outgrown their space can be renovated by cutting back sideshoots to two healthy buds from the main branches.

HAZEL *CORYLUS*

These attractive trees and shrubs have saw-toothed, pointed leaves with soft downy undersides, and pendulous catkins in late winter and early spring. Some types produce edible nuts, which are as attractive to wildlife as to humans, while others are coppiced for fencing and garden supports.

PLANT TYPE Deciduous trees and shrubs
HEIGHT Up to 10m (33ft)
SPREAD Up to 7m (23ft)
FLOWERS ON Last year's stems, in late winter–early spring
LEAF ARRANGEMENT Alternate
WHEN TO PRUNE Late winter–early spring
RENOVATION Yes

Catkins dangle from the bare twisted branches of corkscrew hazel in winter.

HOW THEY GROW

Hazels have an open, airy, upright habit and grow best in chalky (alkaline) soil, in sun or partial shade. *Corylus avellana* 'Pendula' has a weeping habit, while corkscrew hazel (*C.a.* 'Contorta') develops twisted stems. Hanging male catkin flowers appear before the leaves, in late winter, and are typically yellow, but can be pink (for example, *C. maxima* 'Red Filbert' or purple-leaved *C. avellana* 'Rotblättrige Zellernuss').

All varieties of filbert (*C. maxima*) and hazelnut (*C. avellana*) produce edible nuts, which ripen in autumn. Being wind pollinated, they crop better when planted in groups, of at least two varieties, in a square formation. The long straight stems are also harvested.

HOW TO PRUNE

FREE-GROWING TYPES Very little pruning is required. In late winter, shorten wayward shoots back to a main stem or healthy bud, to maintain shape. Thin congested growth and cut out dead and damaged wood. Remove straight-growing suckers from the base of the plant as soon as you see them.

FOR NUTS To get the best crop of nuts, try "brutting". In late summer, bend the new growth on sideshoots in half with your hand until it breaks, but leave the broken section hanging at the end. Then, in late winter, as the catkins are releasing pollen, shorten these broken shoots back to three or so buds. Remove inward-growing stems on the plant, and, if congested, remove up to one-third of old stems back to a stub 2.5cm (1in) from the base.

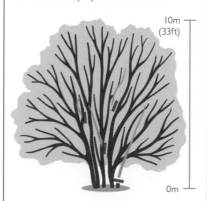

On free-growing hazels, shorten wayward shoots and remove suckers.

COPPICING Hazels can be coppiced every three to five years, but perform better if this is done every five to ten years. Start with a bushy plant with lots of stems instead of a single main stem, and allow it to grow freely for a year or two. Once established, in late winter or early spring, cut back in stages, first cutting the outer stems back to about 30cm (12in) from the base with a pruning saw, felling them all in the same direction. Work towards the centre, pruning every stem in turn. Shorten again to within 5cm (2in) of the ground, creating a "stool" you can cut back to every time you coppice in future years.

RENOVATION PRUNING If a tree or shrub is outgrowing its space, prune hard during late winter or early spring, thinning out any congested growth and cutting back the remaining stems by at least one-half.

Vigorous new shoots develop quickly after hazel has been coppiced.

SMOKE BUSH *COTINUS*

Smoke bushes are grown for their pretty foliage and their clouds of feathery flower plumes in summer. Many varieties produce attractive red or purple leaves, and they can be pruned hard to produce bigger leaves – at the expense of the smoke-puff blooms that give the plant its name.

PLANT TYPE Deciduous small trees and shrubs
HEIGHT Up to 7m (23ft)
SPREAD Up to 5m (16ft)
FLOWERS ON Last year's stems, in summer
LEAF ARRANGEMENT Alternate
WHEN TO PRUNE Late winter–early spring
RENOVATION Yes

Cotinus coggygria **'Notcutt's Variety'** produces a haze of feathery pink flowers.

HOW THEY GROW

When left to grow naturally, smoke bushes tend to develop into neat large domes with quite an open branch structure. They grow well in fertile, moist but well-drained soil, in full sun or partial shade, but for the deepest foliage colour plant in full sun.

Bright-leaved varieties such as 'Grace' and 'Flame' make a gorgeous splash of colour, and can be grown exclusively for their leaves by regularly pruning them hard, or they can be left alone to flower. While the individual flowers are tiny and often an insignificant pinky beige, fading to grey, when massed together in fluffy clusters they create a show-stopping haze. Other popular varieties include *C. coggygria* 'Royal Purple', with its small dark leaves.

HOW TO PRUNE

While smoke bushes tolerate pruning very well, they don't need it regularly. If you wish to reduce the size of your plant, you should bear in mind that the more you trim it the fewer flowers you will get in the summer months.

FREE-GROWING TYPES To enjoy the natural form and size as well as the flowering of these plants, it is best to keep pruning to a minimum. Remove any dead or damaged stems whenever you see them. In late winter, examine the branch structure and if any stems are rubbing remove one. As plants mature, cut out any inward-growing stems and trim overlong new shoots back to an outward-facing bud within the body of the plant, or else right back to the main stem. You can also remove lower branches to allow underplanting to flourish below your smoke bush.

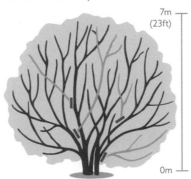

7m (23ft)

0m

Remove any branch crossing the centre, for a healthy, free-growing bush.

FOR LARGE LEAVES To grow smoke bush as a foliage feature plant, allow a newly planted young shrub to grow freely for the first two or three years. Then prune it hard every year, in late winter to early spring, cutting all its stems right back to stumps about 20cm (8in) above the ground. Repeat annually. This will keep the shrub at a height of about 1m (3ft), and it will produce significantly larger leaves on vigorous fresh shoots. You can gradually raise the height of the stumps to about 60cm (24in) if you want the foliage to be held up higher than small plants growing beneath, such as in a border scheme.

RENOVATION PRUNING Use the same technique as for large leaves (*see above*) to renovate an old smoke bush, after which you can either let it grow out again naturally, or continue to prune it back hard every winter for leaf interest.

Hard pruning can be used to renovate a smoke bush, or be practised every year.

COTONEASTER *COTONEASTER*

Cotoneasters are hardy, reliable, easy-care plants producing creamy white flowers in late spring and summer. These blooms are popular with pollinating insects, and are followed by a profusion of bright berries in red, orange, and yellow. Many of the deciduous types have good autumn leaf colour.

PLANT TYPE Deciduous and evergreen trees and shrubs
HEIGHT Up to 10m (33ft)
SPREAD Up to 4m (13ft)
FLOWERS ON Last year's and current year's stems, in late spring–summer
LEAF ARRANGEMENT Alternate
WHEN TO PRUNE Deciduous and wall shrubs in late winter; evergreens in winter or after flowering; hedges after flowering
RENOVATION Yes

HOW THEY GROW

The growth habit varies widely among deciduous and evergreen cotoneasters. Some such as *C.* 'Cornubia' make good small trees, while others such as *C.* 'Rothschildianus' or *C. salicifolius* 'Pink Champagne' are more suitable for growing as informal bushy shrubs. Several like *C. franchetii* are good for hedging, and those with a creeping prostrate habit such as evergreen *C. salicifolius* 'Gnom' make excellent ground-cover plants, especially in tricky or inaccessible spots like a sloping bank or an area of poor soil.

Cotoneasters grow happily in most soil types other than wet soil. Plant deciduous types in full sun, while evergreens grow in sun or partial shade as long as there is no cold wind.

A cotoneaster in autumn, weighed down by berries, is a magnet for birds.

HOW TO WALL-TRAIN

Fishbone or herringbone cotoneaster (*C. horizontalis*) is popular as a wall shrub because of its flat growth. Attach the stems to nails or a wire system on the wall, and tie them in as they grow. Trim back shoots growing into or away from the wall in late winter. This species is now listed as an invasive species in some territories, because birds spread its seed – gardeners are therefore encouraged to be careful when disposing of material after pruning.

HOW TO PRUNE

Although cotoneasters tolerate pruning, they rarely need it – their natural form of spreading branches is part of their charm. Prostrate

Tie stems of fishbone cotoneaster together for better coverage of a wall.

10m (33ft)

0m

Cutting out bare and diseased branches is all that is needed when pruning.

cotoneasters are best left to their own devices, unless you need to keep them within bounds. Other types, particularly *C. × watereri*, are extremely vigorous and need curbing so they do not take over the garden. Varieties used as hedging also require more regular pruning – usually a trim every year after flowering.

DECIDUOUS TYPES In late winter, prune out damaged or diseased shoots, and cut back any old stems that are bare at the base.

EVERGREEN TYPES In late winter or after flowering, shorten any wayward shoots that are spoiling the shape of the plant, cutting back to a main stem. Thin out older stems that are crowding the centre of the plant.

RENOVATION PRUNING Evergreen types of cotoneaster usually tolerate being cut back hard to their desired size in one session, in winter. Renovation of deciduous types should be carried out in stages over a few years.

HAWTHORN *CRATAEGUS*

These hardy trees and shrubs are particularly valuable for wildlife, and for planting in exposed or coastal gardens. Their flat clusters of scented, white or pink flowers are borne in spring. They are followed by typically red but sometimes black or orange fruit, or "haws", in autumn.

PLANT TYPE Deciduous trees and shrubs
HEIGHT Up to 12m (39ft)
SPREAD Up to 6m (20ft)
FLOWERS ON Last year's stems, in spring
LEAF ARRANGEMENT Alternate
WHEN TO PRUNE Free-growing specimens in winter or early spring; hedges after flowering and in autumn
RENOVATION Yes

HOW THEY GROW

The rounded to spreading habit with quite a dense branching structure means that hawthorn is suitable to grow as a hedge, multi-stemmed shrub, and single-stemmed tree. It is tough, vigorous, and easy to grow, doing well in a wide range of soil conditions in full sun or partial shade. Free-growing shrubs and trees should be allowed to reach their potential. Therefore, when choosing where to plant, bear in mind your plant's eventual height and spread, and the access you will need around its thorny branches.

Common hawthorn, also known as May tree (*C. monogyna*), is a great choice for barrier hedges and wildlife-friendly gardens, while earlier-flowering English or Midland hawthorn (*C. laevigata*) cultivars such as dark pink, double-blossomed 'Paul's Scarlet' work well as specimen trees.

HOW TO PRUNE

FREE-GROWING TYPES Hawthorns require only minimal pruning once established. Cut out any crossing or

Common hawthorn carries highly scented, white blossom in late spring.

By autumn, common hawthorn flowers have developed into red haws.

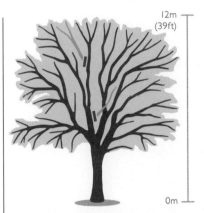

12m (39ft)

0m

On established trees, prune only crossing branches and damaged wood.

misshapen branches as well as any dead and damaged wood, in winter or early spring, to create a well-shaped plant.

HEDGES Trim with shears or secateurs after flowering and in autumn, to keep a neat shape. Make sure young birds have flown the nest before you do the first prune of the year.

RENOVATION PRUNING If it has grown too large for the allocated space, hawthorn can be pruned very hard (almost down to ground level) in winter to bring it back under control. This also prevents it from becoming leggy, and makes for a more compact, bushy plant. When renovation pruning, wear gauntlets and eye protection to avoid injury from the spines while you work your way from the outer shoots inwards, in sections.

DAPHNE *DAPHNE*

These coveted shrubs can be tricky to grow but are worth the trouble for their intensely fragrant flowers. Some types bloom right in the dead of winter, so should be planted near a door or gate so that you can enjoy the heady scent as you pass by during the colder months of the year.

PLANT TYPE Deciduous, semi-evergreen, and evergreen shrubs
HEIGHT Up to 2m (6½ft)
SPREAD Up to 1.5m (5ft)
FLOWERS ON Last year's stems, in late winter, spring, summer, and autumn
LEAF ARRANGEMENT Alternate, rarely opposite
WHEN TO PRUNE Early summer
RENOVATION No

HOW THEY GROW

Many daphnes are upright and bushy, but some have a prostrate or spreading habit. Most are well-behaved, slow-growing plants that do not become very large, so are suitable for even small gardens. Being woodland shrubs, they do best in dappled shade, though they can cope with sun. They prefer slightly acid to slight alkaline soil that is fertile, moist but well-drained. They hate being waterlogged, and drought conditions. All parts of the plant are highly toxic if eaten, and protective gear should be worn when handling as they produce sap irritating to the skin and eyes.

Daphnes can be difficult to establish and must not be moved once planted, so choose their position carefully before planting. They have a reputation for being fussy and are prone to dieback as well as many other problems and diseases like root rot, leaf spot, and viruses. However, despite these challenges, it is their exceptional, sweetly perfumed flowers, often produced at a time of year when little else is happening in the garden, that make daphnes worth the work.

Daphne bholua 'Jacqueline Postill' is a popular, mostly evergreen, medium-sized shrub that grows to about 2m (6½ft). It bears leathery green leaves

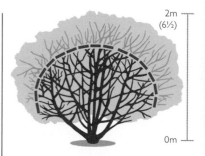

Prune *Daphne bholua* 'Jacqueline Postill' to shape in early summer.

and terminal clusters of scented pink flowers in late winter, followed by black berries. *Daphne odora* 'Aureomarginata' has variegated leaves of green with a yellow edge. *Daphne mezereum* is a good choice for an exposed garden, while *D. laureola* tolerates deep shade, even under trees.

HOW TO PRUNE

Daphnes do not require or like to be pruned, and will not recover from being cut back hard. Some types, such as *D.* × *burkwoodii* and *D. cneorum*, benefit from being left alone, except for having dead and diseased wood cut back to healthy growth where necessary; do this immediately after flowering.

Unlike other daphne varieties, *D. bholua* 'Jacqueline Postill' can develop long leggy growth, in spring. To create a neater, bushier, more compact shrub, trim the new shoots back by about 15cm (6in) with clean secateurs, cutting at an angle just above a set of leaves.

***Daphne mezereum* f. *alba* 'Bowles's Variety'** flowers in late winter.

Shorten the fresh spring growth on *Daphne bholua* 'Jacqueline Postill'.

DEUTZIA *DEUTZIA*

Clusters of starry, white or pink, sometimes scented flowers almost smother these easy-to-grow, deciduous shrubs from mid-spring to midsummer. Some develop attractively peeling bark as the stems age. Most have oval-shaped leaves, but some are narrow and willow-like.

PLANT TYPE Deciduous shrubs
HEIGHT Up to 3m (10ft)
SPREAD Up to 2m (6½ft)
FLOWERS ON Last year's stems, in mid-spring–midsummer
LEAF ARRANGEMENT Opposite
WHEN TO PRUNE After flowering
RENOVATION Yes

Flower clusters smother *Deutzia × elegantissima* during late spring.

HOW THEY GROW

Deutzias typically have an upright graceful habit. The larger varieties make excellent specimen plants, but most work well in a shrub or mixed border. Left to grow freely, some deutzias reach 3m (10ft) high and 2m (6½ft) across, as is the case with *D. × magnifica* 'Staphyleoides', a vigorous shrub with arching branches. If you have a small garden, it would be better to choose one of the dwarf varieties, such as *D. gracilis* 'Nikko', which reaches only 0.6m (2ft) high and 1.2m (4ft) across.

They grow best in loamy fertile soil that is not too dry, but will tolerate limey or poor soil. Most prefer full sun, while some grow well in partial shade. In cold areas prone to sharp frosts in winter, grow less hardy types such as *D. setchuenensis* against a sheltered wall or surrounded by other trees and shrubs. Notable species for scented blooms are *D. gracilis* and *D. scabra*, while D. × *hybrida* 'Strawberry Fields' combines a soft sweet fragrance with pretty pink flowers. *Deutzia pulchra* offers cinnamon-coloured, peeling bark for interest in winter.

HOW TO PRUNE

Once established, free-growing deutzia benefits from an occasional tidy up; regular pruning is not necessary. To encourage more and bigger flowers and new growth from the base, you can remove up to one-quarter of the oldest stems right back to ground level. Do this after flowering, every two or three years. At the same time, cut flowered stems that are getting woody back to just above strong new shoots growing out from farther down the stem.

RENOVATION PRUNING Prune deutzias back hard to keep them within their allotted space, for example when grown in a border. In winter, cut all stems down to within 20cm (8in) of the base or, for a less drastic renovation, to branches within 45cm (18in) of the base. Do not expect flowers the next season.

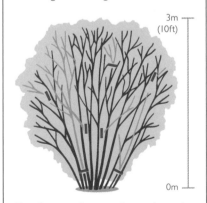

3m (10ft)

0m

Cut flowered stems that are becoming woody back to healthy shoots.

Remove up to one-quarter of old stems at the base, to encourage growth.

ELAEAGNUS *ELAEAGNUS*

Although elaeagnus produces pretty, sometimes scented flowers and edible berries, it is grown mainly for its ornamental foliage in green, silver, or variegated patterns. This is a great plant for exposed and coastal areas because it is tough, drought-tolerant, and fast-growing.

PLANT TYPE Deciduous and evergreen small trees and shrubs
HEIGHT Up to 6m (20ft)
SPREAD Up to 5m (16ft)
FLOWERS ON Last year's and current year's stems, in late spring–autumn
LEAF ARRANGEMENT Alternate
WHEN TO PRUNE Deciduous types after flowering; evergreens in mid–late spring; hedges in late summer–early autumn
RENOVATION Yes

HOW THEY GROW

Elaeagnus is hardy, easy to grow, and provides year-round interest. The deciduous types are usually spreading and open in habit, with arching stems, while the evergreen types are typically denser and bushier in form. Grow elaeagnus in fertile, well-drained soil that is acid to slightly alkaline. It tolerates drier soils and coastal winds, but the leaves may become yellow when grown on poor chalky soil. The most intense colour on the variegated forms is obtained by planting them in full sun, while evergreens and those with plain green leaves also grow in partial shade.

All make good specimen shrubs but some types are used for hedging, too. Commonly grown varieties include: oleaster (*E. angustifolia*), a deciduous type that can grow into a small tree to 6m (20ft) high; and shrubby evergreen *E. × submacrophylla* cultivars, which can reach 4m (13ft).

HOW TO PRUNE

Elaeagnus does not require regular pruning but can be trimmed to keep in shape or to size. Remove any dead, diseased, and crossing branches, and trim those that spoil the natural shape back to the main form of the bush.

EARLY-FLOWERING DECIDUOUS TYPES

After flowering, trim flowered shoots back to stronger buds lower down.

HEDGES Elaeagnus works best as an informal hedge – that is, so that it does not have straight sides and top as you would with a formal privet (*Ligustrum*) or beech (*Fagus*) hedge. During late summer to early autumn, after young

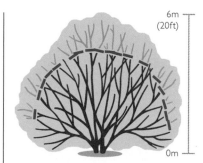

6m (20ft)

0m

On early-blooming deciduous types, shorten the flowered shoots.

birds have flown the nest, trim your hedge, clipping it to mimic the natural growth habit as much as possible.

VARIEGATED TYPES Occasionally, variegated types of elaeagnus throw out a shoot with plain green leaves (returning to its original form). As soon as any are spotted, cut them out at the point from where they are growing, to prevent the whole plant "reverting".

RENOVATION PRUNING Most deciduous elaeagnus should respond to hard pruning. The whole plant can be cut down to your desired size and even back to 15–30cm (6–12in) above ground level, if necessary. Alternatively, over two or three years, cut back shoots to strong growth lower down, and remove one in three stems, starting with the oldest, to promote replacement growth. Prune evergreen *E. × submacrophylla* back hard into old wood if outgrowing its space.

Elaeagnus **'Quicksilver'** is a deciduous type, with silver, willow-like leaves.

Elaeagnus x submacrophylla **'Limelight'** is often used for hedging.

HEATH *ERICA*

Heaths bloom at various times throughout the year, producing masses of tiny, usually bell-shaped flowers ranging in shades of red and pink to white. In some cultivars, the small, tightly curled, needle-like leaves are tinted with red or gold, or colour up in cold weather.

PLANT TYPE Evergreen shrubs
HEIGHT Up to 6m (20ft)
SPREAD Up to 3m (10ft)
FLOWERS ON Last year's and current year's stems, in spring, summer, autumn, or winter
LEAF ARRANGEMENT In whorls, rarely opposite
WHEN TO PRUNE After flowering
RENOVATION Partial

Erica carnea 'Rosalie' has bronze-green foliage and pink blooms in winter.

HOW THEY GROW

These dense bushy plants are good for suppressing weeds, and are also a great choice for coastal areas because they can tolerate salt spray. Hardy prostrate cultivars, like those of low-spreading winter heath (*E. carnea*), make good ground cover and work well in rock gardens when planted among conifers. Taller upright types such as tree heath (*E. arborea*), which can grow up to 6m (20ft) high, are excellent specimen plants in borders.

Heaths prefer an open position in full sun, but a few grow in partial shade. All develop best in moist but well-drained, acidic soil; however, some of the winter- and spring-flowering types

Tree heath *Erica arborea* 'Estrella Gold' grows up to 1.2m (4ft) high.

such as *E. carnea* and *E. × darleyensis* tolerate slightly alkaline soil, as will summer-flowering *E. manipuliflora*, *E. terminalis*, and *E. vagans*.

HOW TO PRUNE

When pruned every year, heaths remain compact and bushy; otherwise, they can become straggly over time, and bare and woody in the centre, and won't flower well.

For their regular annual trim, after flowering, cut back flowered shoots to about 2.5cm (1in) above old growth – known as "staying in the green" – using secateurs or a pair of hedging shears. Remove any dead or damaged stems and clip out any wayward shoots.

30cm (12in)

0cm

Trim back spent flower spikes after flowering, to keep the plant bushy.

RENOVATION PRUNING No heath, apart from *E. arborea*, responds to being hard pruned because they do not regenerate from old wood, so if plants have grown bare at the base or leggy it is best to dig them up and replace with new plants. *Erica arborea*, on the other hand, tolerates being cut back hard into old wood, down to around 9cm (3½in) from ground level, in mid-spring.

Long floral spikes adorn *Erica × darleyensis* f. *albiflora* 'White Perfection'.

GUM *EUCALYPTUS*

Gums are distinctive plants grown for their handsome, often aromatic foliage and ornamental bark. Their leaves are usually mid- or grey-green and leathery, sometimes looking like silver coins while young. Their bark is smooth and white, flaking or striped, or in shades of tawny brown.

PLANT TYPE Evergreen trees and shrubs
HEIGHT Up to 50m (165ft)
SPREAD Up to 35m (115ft)
FLOWERS ON Last year's stems, in spring or summer–autumn
LEAF ARRANGEMENT Opposite when young; alternate when mature
WHEN TO PRUNE Early spring
RENOVATION Yes

HOW THEY GROW

Grow in fertile, neutral to slightly acidic soil that does not easily dry out, preferably in a sunny spot with shelter from cold winds, because branches can break quite easily. In more exposed areas, apply a deep mulch before winter. *Eucalyptus rodwayi* and *E. aggregata* grow in damp and boggy areas, and snow gum (*E. pauciflora* subsp. *niphophila*) is particularly hardy.

Cider gum (*E. gunnii*) and its cultivars usually grow enormous, but *E.g.* France Bleu and *E.g.* Silverana can be grown in a border. *Eucalyptus kybeanensis* is another good choice for a small garden, reaching just 4m (13ft) if it is left to grow unpruned.

Gums develop eye-catching, grey-green foliage and pale brown or white trunks.

Snow gum offers great winter interest with its pale marbled bark.

HOW TO TRAIN

One way to appreciate the often beautifully coloured bark is to train your gum tree with an open centre and to restrict the number of stems to three or four, to create a multi-stemmed tree. While the tree is young, tie each stem with strong string and then secure each string to a brick or large stone. This weighs down the stems and keeps the centre of the tree open, so you can appreciate the colourful bark more. After a season or two, the stems will be trained so the ties can be removed.

HOW TO PRUNE

Prune gums in early spring, just as new growth begins and after the danger of late frost has passed in cold regions. Remove crossing shoots and dead, diseased, and damaged growth back to healthy growth. You can also tip-prune in midsummer if necessary, to encourage bushiness.

RENOVATION PRUNING Gums, especially cider gum, can make very large trees, so to grow them as small bushy shrubs with fresh young foliage, or to restrict their growth, you can coppice or pollard them every year, or once every two or three years. After a gum has been pruned heavily, feed with high-potash fertilizer in mid-spring.

To coppice, cut all stems down to within 8cm (3in) of ground level in winter or early spring. New shoots will quickly grow again.

To pollard a gum tree, cut the shoots back close to the main stem so there develops a head on a clear tree trunk, usually topped at about 1m (3ft). Each year, cut back the new growth close to the main stem in early spring.

50m (165ft)

0m

Remove crossing and inward-growing shoots, and any dead or damaged growth.

EUCRYPHIA *EUCRYPHIA*

This group of mostly evergreen trees and shrubs are valued most for their late, often fragrant, bowl-shaped flowers in white or pink, with fluffy stamens in the centres. The blossom, which attracts bees for its nectar, is borne on established plants from summer to early autumn.

PLANT TYPE Mainly evergreen trees and shrubs
HEIGHT Up to 15m (49ft)
SPREAD 8m (26ft)
FLOWERS ON Last year's and current year's stems, in summer—early autumn
LEAF ARRANGEMENT Opposite
WHEN TO PRUNE Mid-spring; after flowering for *E. lucida*
RENOVATION No

Evergreen *Eucryphia x nymansensis* is smothered in show-stopping blooms during late summer.

15m (49ft)

0m

Clip back any overlong shoots to maintain the natural form of the plant.

HOW THEY GROW

Eucryphias have an upright or columnar growth habit, and usually develop as small or medium-sized trees, apart from *E. milliganii*, which becomes quite large. Plant in fertile, moist but well-drained soil that is neutral to acid. Exceptions are *E. cordifolia* and *E. × nymansensis* 'Nymansay', which tolerate alkaline soil. Position with the top-growth in full sun and the base in shade, because they like their roots kept cool and moist. It can help to grow other plants around eucryphias or to cover the soil with a thick mulch of organic matter, which also shelters them from cold drying winds. Though mostly evergreen, in colder climates some act as semi-evergreens or deciduous plants. The hardiest cultivars are *E. × nymansensis* 'Nymansay' and *E. glutinosa*.

HOW TO PRUNE

Prune eucryphias minimally. Remove crossing and dead wood, and lightly trim back any shoots that spoil the shape of the plant in mid-spring (or, for earlier-flowering *E. lucida*, after flowering). Be careful not to overprune, so you do not lose flowers. Keep large specimens to a reasonable size by regular trimming, but only if necessary.

Eucryphia x intermedia 'Rostrevor' has white flowers and a columnar habit.

SPINDLE TREE *Euonymus*

The main attractions of these trees and shrubs are colourful foliage and, sometimes, unusual decorative fruits that split to reveal seeds in a contrasting colour. Deciduous types develop fiery autumn leaves, while some evergreens have bright variegation that provides interest year-round.

PLANT TYPE Deciduous and evergreen trees and shrubs
HEIGHT Up to 10m (33ft)
SPREAD Up to 8m (26ft)
FLOWERS ON Last year's stems, in spring–early summer
LEAF ARRANGEMENT Mainly opposite
WHEN TO PRUNE Deciduous in late winter–early spring; evergreens in spring and late summer
RENOVATION Yes

HOW THEY GROW

Spindle trees naturally have a bushy habit, particularly the evergreen ones, which are sometimes grown as hedging. Deciduous types, although dense, create a slightly more open habit. Prostrate forms like *E. fortunei* 'Emerald 'n' Gold' and *E.f.* 'Kewensis' form a dense mat of growth, which makes them ideal as ground-cover plants to suppress weeds. *Euonymus alatus* 'Compactus' is a small deciduous shrub that reaches only 1m (3ft). Popular cultivars for striking seed pods and autumn foliage include *E. europaeus* 'Red Cascade' and *E. hamiltonianus*.

Spindle trees grow in any well-drained soil in full sun or partial shade. Those with variegated foliage, such as *E. fortunei* 'Silver Queen', will brighten up a dull corner of the garden, but in shady areas the colour may not be as intense. Those grown in full sun may need moister conditions to thrive, though deciduous types tolerate drier soil. Plant evergreens in a sheltered position out of cold drying winds.

HOW TO PRUNE

DECIDUOUS TYPES Minimal pruning is required. In late winter to early spring, remove dead, damaged, and diseased wood as well as misplaced and crossing branches, to maintain a good shape and a healthy framework.

EVERGREEN TYPES Prune in spring, taking off flowerheads with shears and shortening any shoots that spoil the shape of the plant back to healthy buds.

Evergreens may need a light trim again in late summer just to keep them looking tidy. Watch out for any plain green-leaved shoots on variegated types; if you spot any, immediately cut them back to variegated growth.

RENOVATION PRUNING Neglected deciduous types usually respond well to being pruned hard by cutting all stems back to a low framework or by shortening them to 30–60cm (12–24in) from the ground in late winter or early spring. Evergreens can also be pruned right down to near ground level in mid-spring, just as growth has started. Variegated forms do not respond as well as plain green-leaved types to rejuvenation pruning.

Vibrant red autumn foliage develops on *Euonymus alatus* 'Compactus'.

Cut out green-leaved shoots on variegated plants as soon as you see them.

10m (33ft)

0m

Shorten shoots and deadhead flowers on evergreen types in spring.

SPURGE *EUPHORBIA*

The most cultivated of this very large group of plants are the cacti-like succulent varieties (in tropical areas) and the hardy shrubby and herbaceous types (in temperate climates). Hardy spurges are grown for their architectural form, foliage interest, and distinctive, petal-like bracts.

PLANT TYPE Deciduous and evergreen trees, shrubs, subshrubs, perennials, biennials, annuals, and succulents
HEIGHT Up to 20m (65ft)
SPREAD 3m (10ft)
FLOWERS ON Last year's stems, in spring–summer
LEAF ARRANGEMENT Very variable
WHEN TO PRUNE After flowering and (if necessary) early spring
RENOVATION No

Evergreen *Euphorbia characias* subsp. *wulfenii* flowers in spring.

HOW THEY GROW

Spurges thrive across the world in a range of climates and conditions, meaning that there is a wide variation in size, from just 10cm (4in) to 20m (65ft) high, and in growth habits, from tree-like types to low-spreading varieties. There are large succulents (like *E. canariensis*); upright shrubs (like *E. pulcherrima*); more rounded shrubs (like *E. characias*); clump-forming perennials (like *E. epithymoides*); and prostrate semi-succulents (like *E. myrsinites*) for gardens. The more compact varieties, such as *E. rigida*, which grows 30–60cm (12–24in) high, are ideal for a small garden.

Most hardy spurges are a great choice for coastal gardens. Those that are suitable for growing in a border need full sun and fertile, moist but well-drained soil: for example, the popular perennial *E. characias* subsp. *wulfenii*, with its lime-green spring flowers, or vigorous *E. griffithii* 'Fireglow', with its dark green foliage and orange flowers. Usefully, evergreen

Euphorbia griffithii **'Fireglow'** likes moist soil and spreads quickly.

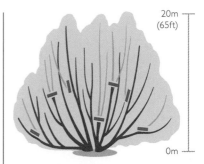

20m (65ft)

0m

Cut flowered shoots back close to the base or to fresh growth further down.

E. amygdaloides subsp. *robbiae* tolerates partial shade and poor dry soil. If you haven't got suitable soil, you can grow spurge in a raised bed or in a container with plenty of grit added to the potting compost.

All parts of spurge are toxic if ingested, and contact with the milky sap can irritate skin, so wear gloves and long sleeves when touching this plant, and especially when pruning it.

HOW TO PRUNE

Most hardy spurges require little attention apart from removing the flowered shoots when they start to fade, cutting to just above a strong shoot or bud lower down the stem. Leave stems that have not flowered as they will do so the following year. In early spring, if necessary, remove dead and damaged stems. Herbaceous perennial types such as *E. griffithii* can be cut back to the ground in autumn.

BEECH *FAGUS*

These magnificent trees are grown for their fine form and wavy-edged foliage. All make wonderful specimens in a large garden, but they are also excellent plants for hedging, pleaching, and topiary. They offer a beneficial habitat for wildlife, which enjoy the nuts inside their prickly seed cases.

PLANT TYPE Deciduous trees
HEIGHT Up to 20m (65ft)
SPREAD Up to 15m (49ft)
FLOWERS ON New stems, in spring
LEAF ARRANGEMENT Alternate
WHEN TO PRUNE Trees in winter; hedges in summer–autumn
RENOVATION Yes

HOW THEY GROW

Beeches are mostly spreading, but some such as *F. sylvatica* 'Pendula' have a weeping or trailing habit, and there are also fastigiate (upright) types like *F. sylvatica* 'Dawyck Purple'.

In most beech trees, the leaves are pale green as they begin to open out and they turn russet-brown or yellow in autumn, but many have dramatic, purple or coppery foliage. When planted densely and clipped as a hedge, beech retains its brown leaves over winter, giving more coverage than other deciduous hedges.

These trees tolerate a wide range of well-drained soils, including chalk, but dislike waterlogged ground. They grow in full sun or partial shade, but for the best-coloured foliage position purple-leaved types in full sun.

HOW TO PRUNE

TREES Large beech trees require little pruning, except where crossing, dead, and damaged branches need to be removed. Do this with a sharp pruning saw, during winter, when the trees are dormant.

HEDGES Cut beech hedges with shears or hedge trimmers, to keep them in shape and to retain the winter leaves.

Fern-leaved beech (*Fagus sylvatica* 'Aspleniifolia') turns golden in autumn.

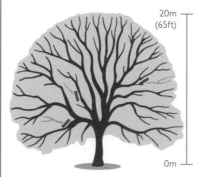

20m (65ft)

0m

Cut out dead, damaged, and crossing branches on established trees.

Common beech (*Fagus sylvatica*) is an excellent choice for informal hedging.

For a dense tight hedge, trim up to three times a year. For a looser look, cut only once a year, in summer.

RENOVATION PRUNING If a hedge has become overgrown, or is too high or wide, you can resort to drastic pruning. Do this in stages, when the hedge is dormant, in winter. In the first year, cut back the top, as well as one of the sides to the desired width. If the hedge shows good recovery, the following winter you can cut back the other side to the required width. Otherwise wait another year before doing so.

FORSYTHIA *FORSYTHIA*

Each spring, these hardy and reliable shrubs are covered in flowers. The bright yellow blooms appear all along the length of the stems, singly or in clusters, before the leaves open. Being versatile, forsythia can be grown as a freestanding plant, a hedge, or be trained up a wall.

PLANT TYPE Deciduous, sometimes semi-evergreen, shrubs
HEIGHT Up to 4m (13ft)
SPREAD Up to 4m (13ft)
FLOWERS ON Last year's stems, in early spring
LEAF ARRANGEMENT Opposite
WHEN TO PRUNE Freestanding and wall-trained shrubs after flowering; hedges after flowering and in late summer
RENOVATION Yes

Forsythia × intermedia **'Lynwood Variety'** flowers in early spring.

HOW THEY GROW

Most forsythias are medium-sized, bushy or upright, deciduous shrubs, but a few are semi-evergreen. They are all low-maintenance plants, but they can develop vigorous, long, leggy, arching stems, so need to be kept pruned if space is tight in your garden. The bright green foliage grows densely, making these shrubs excellent for screening or hedges. Forsythia is also tolerant of urban air pollution and salt in coastal areas, and can be used to help retain soil on sloping sites.

Popular garden cultivars include *F. × intermedia* 'Arnold Giant', 'Lynwood Variety', and WEEK END.

Plant forsythia in any moist but well-drained soil, in full sun or dappled shade. Their explosion of golden-yellow, four-petalled flowers are carried on bare wood for several weeks in early spring, lighting up the garden and providing a good foil to blue-flowered bulbs like scilla and grape hyacinth (*Muscari*). Sometimes, birds eat the buds before they develop into flowers.

HOW TO TRAIN

Some forsythias such as *F.* 'Beatrix Farrand' can be trained up a wall or fence. Attach wires to the structure, or mount a trellis against it, to offer the plant some support. Tie the main stems in and, as the plant grows, regularly tie in all lateral shoots from these stems. Once a lateral is around 30cm (12in), pinch out its growing tip or prune back by one-third. This encourages sideshoots to develop, and it is these that will produce the flowers. After flowering each year, trim sideshoots back to 2–4 buds from the lateral shoot, clip over lightly to neaten the shape, and remove any inward- or outward-growing shoots.

HOW TO PRUNE

FREE-GROWING AND WALL-TRAINED TYPES
It is not necessary to prune forsythia regularly, but it can grow quite large if not kept under control. If you do decide to prune, do so after flowering in spring, removing weak and crossing growth and cutting back flowered shoots to strong buds lower down the stems. Every few years, to encourage fresh growth, cut out one-quarter to one-third of older stems at the base of the plant.

HEDGES Trim more than once a year if necessary, usually after flowering and again in late summer.

RENOVATION PRUNING Forsythias respond well to hard pruning. To rejuvenate an overgrown or leggy shrub, shorten stems down to about 1m (3ft) above the ground, after flowering. The plant probably will not flower much the following spring. However, if you prefer to retain a spring show, cut back one-half of the stems one year, and the remaining old ones the following year.

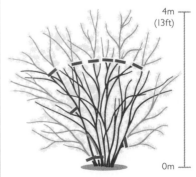

4m (13ft)

0m

Shorten flowered stems and cut out weak and crossing shoots.

FLANNEL BUSH *FREMONTODENDRON*

Fast-growing but short-lived, flannel bushes are large shrubs grown mostly for their big, showy, long-lasting, buttercup-yellow flowers borne prolifically from early summer, with a second flush in, or sometimes continuing into, mid-autumn. They look good when trained against a wall or fence.

PLANT TYPE Evergreen and semi-evergreen shrubs
HEIGHT Up to 6m (20ft)
SPREAD Up to 4m (13ft)
FLOWERS ON Last year's stems, in early summer and autumn
LEAF ARRANGEMENT Alternate
WHEN TO PRUNE Freestanding shrubs in late winter–early spring; wall-trained shrubs after flowering
RENOVATION No

HOW THEY GROW

When grown as freestanding shrubs, flannel bushes can be quite vigorous: for example, F. 'California Glory' can reach 6m (20ft) if left to its own devices. They are not always frost hardy in cool-temperate climates, so might live for only 15 years or so, but despite this they are still worth growing for the show of huge golden blooms they put on each year.

The growth habit is mostly upright at first and then spreading, so flannel bushes make ideal candidates for planting beside or up a south- or west-facing wall, where they can also find the shelter they need from cold drying winds. Place in a sunny dry spot in poor, well-drained soil – the rain shadow of a building is perfect for them. They dislike damp conditions and will rot in wet areas or if overwatered. They tolerate most soil types.

Fremontodendron 'California Glory' has two flushes of blooms each year.

The spreading growth on flannel bush is invaluable for wall training.

The lobed leaves seem hairy or downy but are in fact "scurfy" and carry a dusty substance that will irritate the skin, so always wear gloves when handling this plant.

HOW TO TRAIN

Before planting, attach a system of wires or a trellis on the wall or fence. Choose a central leader stem to tie in and train it in the espalier style (see p.38) or spread out several stems and tie them in, to create a fan style (see p.39). Tie in the lateral shoots that develop from the stem or stems, and let sideshoots grow from these lateral shoots – this is where the flowers will be produced.

HOW TO PRUNE

FREESTANDING TYPES These flannel bushes need only minimal pruning, to retain a good shape and healthy growth. Remove dead, diseased, and damaged shoots back to a healthy stem, and take out any misplaced and crossing growth in late winter or early spring.

WALL-TRAINED TYPES Once established, keep stems to the height required to fit the support, after flowering, and also shorten the lateral shoots to fit appropriately. Clip sideshoots back to 2–4 buds from each lateral shoot and remove inward- and outward-facing growth. At this time you can remove dead wood and crossing shoots, too.

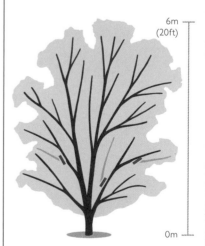
6m (20ft)

0m

Remove dead, damaged, and crossing growth on freestanding shrubs.

FUCHSIA *FUCHSIA*

There are many thousands of fuchsia cultivars, but only a few can survive in temperatures below –5°C (23°F). These hardy fuchsias, with their bell-like, often bicolour blooms, vary from compact bushes to large shrubs, and can also be grown as standards or hedges in the garden.

PLANT TYPE Deciduous and evergreen trees, shrubs, and perennials
HEIGHT Up to 3m (10ft)
SPREAD Up to 2m (6½ft)
FLOWERS ON New stems, in summer–autumn
LEAF ARRANGEMENT Opposite, occasionally in whorls, rarely alternate
WHEN TO PRUNE Spring
RENOVATION Yes

HOW THEY GROW

Fuchsias grow in any moist but well-drained soil in full sun or partial shade. Hardy types are typically deciduous, but some may retain their leaves in climates where winter temperatures stay above 4°C (39°F). While hardy types are more tolerant of frost than other fuschias, many are still borderline, so always plant these in a position sheltered from frost and cold drying winds, when in the ground.

Move container-grown plants into a frost-free place for the colder months so they do not suffer from frozen stems and roots. The stems of even the hardier types may die back over winter in cooler climates, so give them a deep organic mulch in autumn.

Hardy fuchsias range from 40cm (16in) tall to a maximum of about 3m (10ft). Some like *F.* 'Ben Jammin' have a low bushy growth habit, while others such as *F.* 'Lady Boothby' are upright and can be trained against a wall. Larger shrubby types like *F.* 'Hawkshead' make lovely specimen plants. In mild locations, cultivars such as *F.* 'Riccartonii' or *F.* 'Mrs Popple' can be used for a flowering hedge. *Fuchsia* 'Margaret' and *F.* 'Phyllis' are less hardy but very vigorous.

The commonly two-tone, pendent flowers, which are often compared to little ballerinas wearing tutus, come in all sorts of combinations of white and pink to red and purple, and in several forms including single and double. They are followed by edible, if not always tasty, berries.

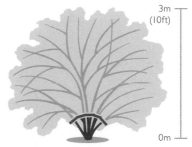
Cut back frost-damaged stems to a low permanent framework.

HOW TO PRUNE

Even the hardier fuchsias may die back completely or partially over winter in colder areas, just as herbaceous perennials do. Cut back dead stems to the base in spring, as growth begins, taking care not to damage the new shoots coming out of the ground. For partial dieback, shorten damaged shoots to just above a healthy bud, or back to a permanent framework about 30cm (12in) from the base.

HEDGES Loosely clip sideshoots back to healthy buds in spring, but keep the form informal with some arching growth left for flowers.

RENOVATION PRUNING In spring, as new growth begins, you can cut an overgrown hardy fuchsia back to the ground to rejuvenate it. To renovate hedges, but retain some cover, cut back every other plant one year, and the remainder the following spring.

Deciduous *Fuchsia* 'Riccartonii' is one of the hardiest fuchsias.

Fuchsia magellanica* var. *molinae bears its dainty flowers in late summer.

SILK TASSEL BUSH *GARRYA*

Eye-catching, extra-long, grey-green male catkins, which appear in midwinter through to early spring, provide a welcome show in the garden during the colder months of the year. Though it can be grown as a freestanding shrub, silk tassel bush is often trained loosely against a wall.

PLANT TYPE Evergreen small trees and shrubs
HEIGHT Up to 4m (13ft)
SPREAD Up to 4m (13ft)
FLOWERS ON Last year's stems, in midwinter—early spring
LEAF ARRANGEMENT Opposite
WHEN TO PRUNE After flowering
RENOVATION Yes

The long, eye-catching, pendulous catkins of *Garrya elliptica* appear in spring.

HOW THEY GROW

Most silk tassel bushes have a dense or bushy, upright growth habit. They offer year-round interest with their leathery, sometimes wavy-edged, green leaves, and are an excellent choice for coastal areas. Grow in full sun or partial shade in any well-drained soil. Although some are frost hardy down to −10°C (14°F), they all benefit from a sheltered position that provides protection from cold winds and frosts, which can damage the leaves and early flowers.

There are male and female forms of silk tassel bush, and the long dangling catkins of the male types are generally more sought-after than the female ones. *Garrya elliptica* 'James Roof' is a popular male cultivar, with beautiful clusters of hanging, silver-green catkins, up to 20cm (8in) long.

HOW TO WALL-TRAIN

Train plants quite loosely against a wall using a system of wires. Attach the stems to the wires via bamboo canes, to form a fan shape, or else create an espalier by training shoots laterally along the wires, off a central leader. Tie in lateral shoots and sideshoots as they grow to fill in the framework.

HOW TO PRUNE

FREESTANDING TYPES These require minimal pruning, but can be trimmed after flowering, while the catkins fade and before new growth is showing. Shorten flowered shoots and cut back any lateral shoots that are sticking out, to neaten the shape while retaining the

Prune back lateral shoots to keep a wall-trained bush in its allotted space.

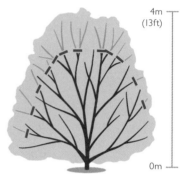

4m (13ft)

0m

Shorten flowered and wayward shoots to retain a neat natural shape.

natural form of the plant. Then cut away any crossing and inward-growing lateral shoots and sideshoots back to the stem, and remove any damaged growth back to a healthy bud.

WALL-TRAINED TYPES Once established, keep silk tassel bush to your preferred height and width with an occasional prune, after flowering. Clip back to the desired size using secateurs, always making your cuts above a healthy bud. Remove any shoots growing out from the wall.

RENOVATION PRUNING Overgrown plants can be rejuvenated over three or four years, after flowering, by cutting back up to one-quarter of the stems each year to a low framework about 30cm (12in) from the base. When new shoots emerge and establish, select the best-placed, strongest ones and thin out badly placed and weak ones.

WITCH HAZEL *HAMAMELIS*

Witch hazels are long-lived, hardy plants valued in the garden for their unusual scented flowers, which spring up, in winter, like little flames along the bare branches before the leaves appear. Many cultivars are also admired for their excellent autumn foliage colour in shades of red, orange, and yellow.

PLANT TYPE Deciduous small trees or shrubs
HEIGHT Up to 4.5m (14½ft)
SPREAD Up to 4m (13ft)
FLOWERS ON Last year's stems, in late winter–early spring
LEAF ARRANGEMENT Alternate
WHEN TO PRUNE After flowering
RENOVATION Partial

HOW THEY GROW

These deciduous shrubs are clump-forming, with a spreading growth habit and arching branches. Being slow-growing, witch hazels take a long time to reach their typical mature size, but do keep their eventual height and spread in mind when planting to allow them the space they need to develop unhindered. They are generally grown as specimen shrubs, and should be placed where the sight and scent of their vibrant, spider-like, yellow, orange, or red blooms can be easily enjoyed in the depths of winter.

Plant in an open, preferably sunny spot in any fertile, moist but well-drained soil that is neutral to acid. Although witch hazels are very hardy, young plants and new growth are susceptible to frost damage, so avoid planting in an area likely to attract frost.

Hamamelis × intermedia 'Pallida' is a popular cultivar bearing yellow flowers with a fruity fragrance, and has orange autumn leaves, while *H. × i.* 'Aphrodite' produces orange flowers and yellow autumn leaves, and *H. × i.* 'Rubin' carries dark red blooms.

HOW TO PRUNE

The only pruning witch hazels need is the removal of any dead and damaged growth. However, you may wish to keep a mature plant contained within its allotted space. In this case, after flowering, take out weak and crossing shoots, and cut all the previous season's growth back to two leaf buds.

Always make pruning cuts just above a healthy bud, using sharp secateurs.

Many witch hazels are grafted on to rootstocks, and may send up suckers from below the graft point or beneath the ground. They are easy to identify and therefore remove in autumn, because suckering shoots lose their leaves later than the rest of the plant. Tear or cut away suckers as close to the stem or base as possible.

RENOVATION PRUNING Every year after flowering, for several years, overlarge or poorly shaped witch hazel shrubs and trees can be reduced in stages by cutting back a few old branches to newer growth lower down.

Hamamelis × intermedia **'Aphrodite'** bears large fragrant flowers in winter.

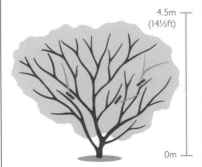

4.5m (14½ft)

0m

After flowering, take out any dead, damaged, and crossing growth.

HEBE *HEBE*

These shrubs and small trees are grown for their year-round foliage interest and rounded forms, as well as their distinctive flower spikes in a range of colours from purple and blue to pink, red, and white. The blooms usually appear from midsummer and can continue all the way through autumn.

PLANT TYPE Evergreen small trees and shrubs
HEIGHT Up to 2.5m (8ft)
SPREAD Up to 2.5m (8ft)
FLOWERS ON Last year's and current year's stems, in summer–autumn
LEAF ARRANGEMENT Opposite, sometimes 2- or 4-ranked
WHEN TO PRUNE Spring
RENOVATION Partial

Hebe cupressoides **'Boughton Dome'** forms a dense green mound.

HOW THEY GROW

Hebes range in size from dwarf bushes to medium-sized shrubs and small trees. Most have a dense mounding habit, forming a dome shape, but a few are spreading, upright, or open-centred. There are hebes available to suit a broad range of planting needs, from coastal areas and rock gardens, containers, and borders to hedging and ground cover. There are low, small-leaved types (like *H. rakaiensis*) and the yellow, conifer-like *H. ochracea* 'James Stirling'; big, upright, large-leaved types like *H.* 'Great Orme' as well as variegated types (like *H.* 'Silver Queen');

and some that change to shades of red and purple during the winter, like *H.* 'Blue Gem'.

However, most have green to grey foliage year-round and provide welcome evergreen structure in the garden throughout the seasons. The flowers can be stubby spikes or long bottlebrushes, and are produced over a long season, with *H.* 'Autumn Glory' in particular providing a show from midsummer right through to late autumn. Deadhead regularly for the best results.

Hebes hate being waterlogged, so plant in well-drained soil that is close to neutral. They prefer growing in the open in full sun, but tolerate light shade. Position in a sheltered place to protect them from frosts and cold winds.

HOW TO PRUNE

Hebes do not usually require much pruning, but you can help to keep the growth compact with an all-over trim

Prune back frost-damaged growth in early spring, with shears.

in early spring, when new buds become visible. For small varieties, clip back just the top 5cm (2in) of growth; for larger plants, you could trim off up to 15cm (6in) – but in either case make sure to leave at least two buds on each stem. Also prune any frost-damaged and wayward stems that ruin the shape of the shrub, back to healthy growth.

RENOVATION PRUNING If your hebe has leggy bare stems with new growth only at the top, it might be best to replace it, because these plants will not respond to being cut back into old wood. However, if there are strong-growing shoots down the stems, you could attempt a partial renovation to reduce the size of an overgrown shrub. Do this in two stages. In spring, trim back the whole plant, to encourage buds to break lower down the stems. If this succeeds, you can trim back further into the plant in late summer, but always leaving at least two buds on each stem.

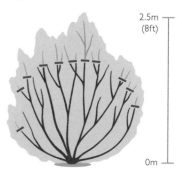

2.5m (8ft)

0m

Trim the top-growth all over to create a neat compact form on your plant.

IVY  *HEDERA*

When the right cultivar is chosen, ivy can be an excellent garden plant, with its glossy evergreen leaves, its ability to climb or be ground cover, and its tolerance of dry shade. In its adult form, its flowers and berries are also an important food source for pollinators and birds.

PLANT TYPE Evergreen climbers
HEIGHT Up to 10m (33ft)
SPREAD Up to 4m (13ft)
FLOWERS ON Last year's stems, in autumn
LEAF ARRANGEMENT Alternate on juvenile forms; spiral on adult forms
WHEN TO PRUNE Spring
RENOVATION Yes

Juvenile forms of ivy have lobed leaves and climbing stems with clinging roots.

HOW THEY GROW

Ivies are vigorous plants with two distinct phases of growth. In the juvenile stage, the climbing or creeping stems have lobed leaves as well as adhesive aerial roots (see p.35) to help them cling to the soil or surfaces. In the adult stage, which can take ten years or more to reach, they change into shrubs with unlobed leaves, and begin to produce umbels of yellow-green flowers in autumn and globed clusters of black berries in winter. Somewhat unusually, *H. helix* 'Arborescens' has exclusively adult growth.

Among the attractions of ivy are that it can used as a climber to quickly clothe a structure or cover an eyesore, or as ground cover, and it is one of the most shade-tolerant plants available. It also grows in full sun, and variegated types like *H. algeriensis* 'Gloire de Marengo'

Ivy can be removed gently from tree bark although it does not harm the tree.

colour better with more light. It prefers rich and fertile, moist but well-drained soil that is slightly alkaline, yet tolerates most situations. The hardiest varieties are cultivars of *H. helix* and *H. colchica*.

Many myths abide about ivy growing on trees and walls. However, it has been shown that they do not harm trees they grow on, and will not damage sound masonry – though they will take advantage of any crack or hole that already exists. If this is of concern, then choose a twining type that needs support, such as *H. colchica* 'Dentata Variegata'.

All parts of ivy are toxic to ingest, and the sap can irritate skin, so wear protection when handling.

HOW TO PRUNE

To remove a mass of climbing ivy from a wall or tree, cut the stems near the soil or low down where they meet

the structure, and wait until they wither and die before removing the stems. Resist the urge to pull them straight off, because this can damage the trunk or paintwork. Look out for plain green shoots on variegated ivies, and prune out as soon as you see them.

JUVENILE STAGE Keep climbing stems to size by cutting back excess growth in mid-spring.

ADULT STAGE To prune an ivy that is grown as an informal hedge or shrub for wildlife, trim back growth to the desired size in mid-spring. Once stems have reached the adult stage, no matter how hard you prune them they will produce fresh adult-stage growth and will not revert to juvenile-stage stems and leaves.

RENOVATION PRUNING Ivy is extremely resilient and responds to hard pruning. If necessary, cut right back into old wood in early spring, to reduce the plant to your preferred size.

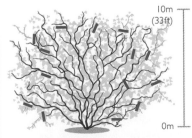

Keep climbing stems in check by clipping them back in mid-spring.

HELICHRYSUM *HELICHRYSUM*

Striking, often aromatic foliage is a hallmark of helichrysum, which includes the pungent curry plant (*H. italicum* subsp. *serotinum*) and fragrant liquorice plant (*H. petiolare*). Mounding types give structure year-round, and the silver-grey leaves make a cool contrast to pastel shades in the border.

PLANT TYPE Evergreen shrubs, subshrubs, perennials, and annuals
HEIGHT Up to 1.2m (4ft)
SPREAD Up to 1.2m (4ft)
FLOWERS ON Last year's and current year's stems, in late spring or summer—autumn
LEAF ARRANGEMENT Alternate; sometimes opposite or rosette
WHEN TO PRUNE Spring and summer
RENOVATION No

Curry plant produces its domed clusters of small flowers in summer.

The foliage on liquorice plants is soft and covered with very fine hairs.

HOW THEY GROW

Helichrysums vary from small evergreen shrubs and subshrubs to herbaceous subshrubs, perennials, or annuals, often with their character depending on the climate where they are grown. They range from prostrate types that grow to just 5cm (2in) tall, to cushion-forming and bushy mounding shrubs reaching up to 1.2m (4ft). The leaves can be hairy and offer the same kind of soft grey foliage interest as lavender (*Lavandula*). Helichrysum grows in any well-drained soil as long as it is in a dry and sunny spot, sheltered from cold winter winds and wet. It is useful for coastal gardens as well as drought-tolerant schemes.

Curry plant is one of the toughest helichrysums, being hardy down to −5°C (23°F), though it can still suffer frost damage in cold winters. With its long narrow leaves that smell intensely of curry, it is a popular choice for sensory gardens. In order to enjoy the attractiveness of its foliage, some gardeners snip off the distracting flowerheads as they develop on this plant. *Helichrysum stoechas* and its cultivars, including *H.s.* 'White Barn', have a similar frost hardiness.

Liquorice plant is a more tender, trailing, non-edible plant with heart-shaped, woolly, grey leaves. It is often treated as an annual in cool-temperate climates, where it is grown in a container or hanging basket.

A related plant – strawflower or everlasting flower – was previously known as a helichrysum but is now called *Xerochrysum bracteatum*. This is grown for its papery, daisy- and chrysanthemum-like blooms in shades of orange, yellow, pink, red, and white, which are used by many people in dried flower arrangements.

HOW TO PRUNE

Shrubby helichrysums benefit from regular pruning to keep them compact, because they can become straggly if left to grow and flower freely. Clip with topiary shears or secateurs in spring as growth begins, cutting the previous year's growth back to one or two buds. Shorten any frost-damaged shoots and stems back to healthy growth, and remove any leggy weak growth back to near the base of the plant.

Taking off flower buds as soon as they appear will help keep the plant dense, but, if you wish to allow it to flower, deadhead the spent blooms afterwards. A trim all over in summer also helps to maintain the shape.

Helichrysums do not usually respond to being pruned back hard, so overgrown, bare-legged plants are best replaced.

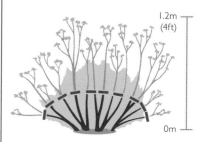

1.2m (4ft)

0m

Cut all the previous year's growth back to one or two buds in spring.

HIBISCUS *HIBISCUS*

Hibiscus are showy shrubs known for their eye-catching, trumpet-shaped flowers in a range of colours from pink and red to blue, yellow, and white, which add a tropical touch to a mixed border. Hibiscus shrubs vary in hardiness, but all require some shelter in cool-temperate climates.

PLANT TYPE Evergreen and deciduous shrubs
HEIGHT Up to 5m (16ft)
SPREAD Up to 3m (10ft)
FLOWERS ON Current year's stems, in summer–autumn
LEAF ARRANGEMENT Alternate
WHEN TO PRUNE Late spring
RENOVATION Yes

HOW THEY GROW

Deciduous types can be grown outside year-round, while evergreen hibiscus are better in containers that are moved into a frost-free place like a greenhouse or conservatory for winter. Rose of Sharon (*H. syriacus*) is an upright, vase-shaped, hardy deciduous type with dark green, hand-like leaves divided into leaflets and large, showy, funnel-like flowers. Popular cultivars include *H.s.* 'Oiseau Bleu', with unusual, red-centred, blue flowers, and *H.s.* 'Woodbridge', with pink flowers and dark pink centres. They need a long hot summer to flower well and, although frost hardy, grow best in full sun in a sheltered position, which is south- or west-facing. They prefer moist but well-drained soil that is neutral to alkaline. Some can tolerate clay soils, but all hate being waterlogged.

Varieties of tropical hisibcus (*H. rosa-sinensis*), a rounded bushy evergreen, are good for growing in pots so they can be brought indoors before temperatures drop below 10°C (50°F). Place outside in summer, out of direct sunlight.

HOW TO PRUNE

Allow newly planted shrubby hibiscus to establish for a few years without much pruning, just trimming the tips to promote a bushier habit. Once they reach the desired height and spread, you can lightly prune each year in late spring, after the threat of hard frosts has passed, to keep them to size and a tidy shape.

DECIDUOUS TYPES Take out any dead, damaged, crossing, or congested shoots, and trim back weak outer shoots.

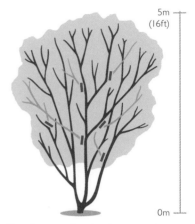

5m (16ft)

0m

Regular pruning involves removing weak, damaged, and crossing growth.

Hibiscus are prone to developing drooping outer branches or to growing lopsided if left to their own devices. Reshape in late spring by cutting back the badly formed stems to the base or to healthy strong vertical growth farther down. Deadhead as the flowers fade.

EVERGREEN TYPES Prune and deadhead as for deciduous types but shoots can also be shortened to keep them to a suitable size for their container.

RENOVATION PRUNING You can attempt to renovate an overgrown specimen completely in late spring by pruning out the oldest stems and cutting back the remainder to within 30cm (12in) of the base. However, if a plant has suffered extensive dieback, it is probably best to replace it.

Hibiscus syriacus **'Oiseau Bleu'** flowers from late summer onwards.

Drooping lower stems can be pruned back to the base in late spring.

HYDRANGEA *HYDRANGEA*

Hydrangeas are elegant, easy-to-grow garden plants, beloved for their long-lasting, showy flowerheads in hues of mostly pink, purple, blue, and white, which brighten up any scheme in late summer. They vary in size, form, leaf, flower shape, and colour, but all enjoy similar growing conditions.

PLANT TYPE Deciduous and evergreen shrubs and climbers
HEIGHT Up to 15m (49ft)
SPREAD Up to 4m (13ft)
FLOWERS ON Last year's or current year's stems, in mid–late summer
LEAF ARRANGEMENT Opposite or in whorls
WHEN TO PRUNE Spring for most hydrangeas; after flowering for climbers
RENOVATION Yes

HOW THEY GROW

Hydrangeas are excellent flowering shrubs for the mixed border, and are often also grown as specimen plants and in containers. There are many different forms, but most garden types are deciduous shrubs that grow to 2m (6½ft) and flower from midsummer.

Some of the most popular are *H. macrophylla* cultivars, which are divided into two groups: lacecaps and mopheads (also called hortensia hydrangeas). Lacecaps, as the name suggests, have flattened flowers with small petals in the centres and frills

TOP TIP BLUE-FLOWERED HYDRANGEAS WILL BLOOM PINK UNLESS GROWN ON ACIDIC SOIL WITH A PH OF AROUND 5, WHICH IS RICH IN THE ALUMINIUM SALTS THAT TRIGGER THE BLUE COLOURING. IF YOU DON'T HAVE ACIDIC SOIL IN YOUR GARDEN, YOU CAN APPLY A BLUING AGENT DURING THE GROWING SEASON. ALTERNATIVELY, TRY ADDING RUSTY NAILS OR TEA LEAVES TO THE SOIL AROUND THE PLANT.

of larger petals around the edge, while mopheads develop domed flowerheads of tightly packed petals. Both types have a rounded growth habit, as does *H. arborescens*, whose popular cultivar 'Annabelle' produces large, white, spherical flowers in summer.

Vigorous *H. paniculata* and its cultivars are more upright and spreading, with cone-shaped flowers,

while *H. quercifolia* has oak-like leaves and grows as a large mound. Taller growing are the climbers, which include deciduous *H. anomala* subsp. *petiolaris* and evergreen *H. seemannii*; the former is fully hardy while the latter is frost hardy to about −5°C (23°F). *Hydrangea serrata* is also frost hardy, and is compact, growing up to 1.2m (4ft), so a good choice for a small garden.

The petals on *Hydrangea paniculata* PINKY-WINKY turn from white to pink.

Hydrangea macrophylla will flower pink on neutral to alkaline soil.

Grow Hydrangea anomala subsp. *petiolaris* up against a wall.

Hydrangea arborescens **'Annabelle'** flowers on new growth each summer.

Although most hydrangeas are fully hardy, they can be damaged by intense sun or cold winds, so they benefit from a sheltered situation with some shade. They grow in any moist but well-drained, fertile soil, but may develop yellow leaves on poor or chalky soil. They do well in clay soils, and you can improve moisture retention with a mulch in spring on any, especially drier, soils. If planted in a container, use ericaceous (lime-free) compost.

HOW TO PRUNE

Some hydrangeas flower on old wood, while others do so on new growth, and this difference affects how they are pruned. The group that bloom on last year's growth includes *H. macrophylla*, *H. quercifolia*, *H. serrata*, and the climbing types (*H. anomala* subsp. *petiolaris* and *H. seemannii*). Types that bloom on the current year's growth include *H. paniculata* and *H. arborescens*. In all cases, as a minimum, damaged shoots should be trimmed back to healthy growth and weak or trailing stems cut back to the ground.

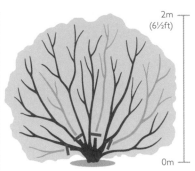

Take out one in three old stems on established *Hydrangea macrophylla*.

TYPES THAT BLOOM ON LAST YEAR'S GROWTH
For *H. aspera* and *H. quercifolia*, the only pruning necessary is of dead growth and wayward stems in spring.

On *H. macrophylla*, the fading blooms of lacecap types should be removed after flowering, to stop the plant wasting energy by setting seed. Cut back to the second set of leaves below the flowerhead. Mophead types benefit from their old flowerheads being retained over winter, to protect the new buds. Take them off in spring by cutting back to the first set of buds below the head. As well as this, on both types, you can

Cut back to the second set of leaves below the flowerhead on lacecap types

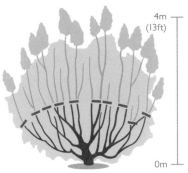

Cut last year's stems on *Hydrangea paniculata* back to a lower set of buds.

remove one-third of the oldest stems right back to the base, on established plants, in spring. Prune *H. serrata* in the same way as *H. macrophylla*.

For climbing types, prune after flowering. Cut back any shoots reaching beyond their bounds, to keep the plant to its allotted space.

TYPES THAT BLOOM ON CURRENT YEAR'S GROWTH
Hydrangea arborescens and *H. paniculata* do not require much pruning beyond removing dead and weak growth, but they will flower more profusely if you cut them back each spring. Shorten last year's stems to a set of healthy buds, and prune any old unproductive stems back to the ground. Alternatively, to encourage larger (though fewer) flower panicles, cut the stems back even farther, to just above the lowest pair of buds. This creates a permanent framework to prune back to every year.

RENOVATION PRUNING Overgrown and neglected *H. macrophylla* shrubs can be rejuvenated by cutting back all stems right to the base in spring. However, they will not produce any flowers in the summer following this hard pruning.

Well-established climbing hydrangeas can be renovated, but do this in stages. Every year for three years, after flowering, remove one-third of the oldest stems, and trim back the remaining ones by one-half, to above strong healthy buds.

ST JOHN'S WORT *HYPERICUM*

Vigorous, low-maintenance St John's wort is an easy-to-grow plant that tolerates most conditions, including full shade and drought. It is grown mainly for its copious, large, yellow flowers. Some types have red autumn foliage, and many also produce pretty, berry-like fruit capsules.

PLANT TYPE Evergreen, semi-evergreen, and deciduous shrubs
HEIGHT Up to 1.5m (5ft)
SPREAD 1.5m (5ft) or more
FLOWERS ON Current year's stems, in summer–autumn
LEAF ARRANGEMENT Opposite, occasionally in whorls
WHEN TO PRUNE Early spring
RENOVATION Yes

HOW THEY GROW

St John's wort varies in size and form from upright shrubs with arching branches to domed bushes and low-spreading, prostrate types. Each growth habit is suitable for different situations, from specimens in a shrub border to hedging and ground cover. Most of the shrubby types are very hardy, although the evergreens may lose their leaves in cold winters; however, they will revive with fresh growth in spring. In frost-prone areas, some shelter might be beneficial. St John's wort tolerates most growing conditions, doing best in partial shade, but some thrive in full shade, and others in full sun. It grows well in almost all soil types, and is drought-tolerant. Some species are renowned for self-seeding and others for spreading, but in general they require little maintenance.

One of the most popular cultivars is *H. × hidcoteense* 'Hidcote', which is a semi-evergreen, dense, and bushy shrub that grows to about 1.2m (4ft), and carries big, saucer-shaped, blazing gold blooms from late summer to autumn. *Hypericum calycinum* is an evergreen that makes a good choice for a site in full shade. It forms excellent ground cover, but is known to spread through runners. *Hypericum × moserianum* 'Tricolor' develops eye-catching, variegated, cream, pink, and green leaves, while *H. × inodorum* 'Elstead' produces large, dark pink fruit capsules after the flowers.

HOW TO PRUNE

St John's wort requires minimal pruning, but to keep the more vigorous varieties in check and retain their dense growth habit it is a good idea to trim them in early spring. Remove any dead,

Each flower on St John's wort has a sunburst of stamens at the centre.

Prune deciduous types back to a low permanent framework each spring.

damaged, or diseased wood back to healthy growth, and take out crossing and inward-growing shoots. Remove leggy stems at their point of origin.

DECIDUOUS TYPES Cut back each year to a low permanent framework.

EVERGREEN AND SEMI-EVERGREEN TYPES Reduce in size by shortening all stems back to healthy buds. Cut *H. calycinum* right back to the ground each year to control its growth.

RENOVATION PRUNING St John's wort becomes straggly if neglected, but overgrown specimens respond to hard pruning and even drastic renovation. In early spring, simply cut all stems down to ground level.

1.5m (5ft)

0m

Trim all over in spring to encourage compact dense growth.

HOLLY *ILEX*

Holly is famed for its glossy evergreen leaves and bunches of cheery red berries, often harvested for festive decorations in winter. It is suitable for hedging, topiary, and freestanding plants. Although known for its prickly, plain-green foliage, there are smooth-leaved and variegated types as well.

PLANT TYPE Evergreen trees and shrubs
HEIGHT Up to 15m (49ft)
SPREAD Up to 5m (16ft)
FLOWERS ON Last year's stems, in spring
LEAF ARRANGEMENT Alternate
WHEN TO PRUNE Freestanding in late winter—early spring; topiary and hedges in spring and summer
RENOVATION Yes

HOW THEY GROW

Hollies vary in size, shape, and habit, from conical or pyramid-shaped trees to columnar and bushy shrubs. Many gardeners use box-leaved or Japanese holly (*I. crenata*) for creating a low-growing hedge up to 1m (3ft) tall. The spiky foliage of types like common holly (*I. aquifolium*) makes them unsuitable for growing in a border, so you might prefer to plant one of the smooth-leaved varieties. Variegated cultivars are also available, with gold-splashed or silver-edged leaves. Some produce yellow or black berries rather than the more normal red ones.

Ilex x meserveae **Blue Angel** is a female holly with the classic red berries.

Most hollies are hardy and tough enough to cope with windy and coastal sites. They grow in any moderately fertile, moist but well-drained soil in sun or partial shade, though variegated types colour better in full sun.

FOR BERRIES You need to plant freestanding male and female types near each other – the berries are produced on a female plant, once its flowers are pollinated by insects that have also visited a male plant. To discover the sex of an established bush, check its flowers: male holly blooms have four stamens, while female flowers have a green berry at the centre. Do not assume sex by the name of the cultivar: for example, *I. × altaclerensis* 'Golden King' is actually a female type, while *I. aquifolium* 'Silver Queen' is male. There is also a self-fertile variety called *I.a.* 'J.C. van Tol'.

HOW TO PRUNE

The more prickly-leaved hollies are clipped, the spinier the leaves become, so wear protection such as gauntlets while pruning. Where possible, clip the stems rather than through the leaves, as the cut foliage will develop brown edges. Remove green-leaved stems on variegated types as soon as you see them.

FREESTANDING TYPES These require little pruning beyond removing all dead and damaged growth, and any crossing shoots, in late winter or early spring.

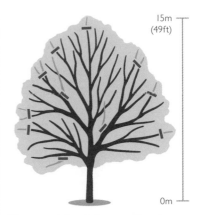

15m (49ft)

0m

Prune out dead, damaged, and crossing shoots on freestanding plants.

HEDGES AND TOPIARY Hollies can be kept to any size or shape with regular trimming, which is what makes them so suitable for formal shapes. Spring is a good time to clip topiary, but you may find that a little-and-often approach also works well, by trimming lightly in spring and then again in mid- and late summer. Informal hedges can be clipped once in summer, while more formal ones may need more than one trim, in spring and summer.

RENOVATION PRUNING Hollies respond well to hard pruning, so overgrown plants or those that have become bare at the bottom and centre can be rejuvenated in late winter by cutting right back into old wood, even down to the base. It can take time for the plant to send up new growth after such a drastic renovation.

JASMINE *JASMINUM*

The white, cream, or yellow flowers of jasmine are star-shaped and scented, spicing up the garden with their heady perfume. These vigorous climbers and wall shrubs will quickly clothe a wall, fence, arch, or trellis with their cheerful blooms and pretty leaves divided into leaflets.

PLANT TYPE Evergreen and deciduous shrubs and climbers
HEIGHT Up to 5m (16ft)
SPREAD Up to 5m (16ft)
FLOWERS ON Last year's and current year's stems, in summer or winter
LEAF ARRANGEMENT Opposite or alternate
WHEN TO PRUNE After flowering
RENOVATION Yes

Common jasmine is a summer-flowering climber with fragrant flowers.

HOW THEY GROW

There are not only climbing forms of jasmine, which twine with tendrils (see p.35), but also shrubbier types with lax arching stems that don't twine but can be trained against a wall. Some are evergreen and others are deciduous; and some are hardy while others are tender and need to be grown indoors or brought in over winter in cool-temperate climates. Most flower in summer, but shrubby deciduous winter jasmine (*J. nudiflorum*) produces its very fragrant, yellow blooms in winter. It thrives in full sun or partial shade.

Common jasmine (*J. officinale*) is a deciduous climber with scented white flowers. It is frost hardy down to about

The bright yellow blooms of winter jasmine appear on bare green stems.

–5°C (23°F) and appreciates a sheltered position in full sun. Both these popular species grow in any fertile, well-drained soil.

The evergreen, white-flowered false or star jasmine (*Trachelospermum jasminoides*) is not a true jasmine, but is often considered one. Prune out awkward growth in spring.

HOW TO TRAIN

Winter jasmine and other shrubby types can be fanned out and tied loosely to a wire system or support on a wall, with new shoots secured as they grow to fill in the gaps. After planting, direct climbers like summer jasmine towards a trellis or other support by

tying their stems on a cane angled towards that support. Tie in new stems as they grow.

HOW TO PRUNE

All jasmine benefits from an annual prune, after flowering. Thin out dead, crowded, crossing, misplaced, or weak stems, then cut back flowered stems to strong sideshoots lower down. Trim the leafy growth of summer jasmine too, to keep it neat and to size.

RENOVATION PRUNING Jasmines tolerate hard pruning. If a plant has grown too large, you can cut back outer growth with shears before taking out the oldest stems, and reducing the length of the others to the desired size, with secateurs. Alternatively, if you don't mind losing flowers for a few years, chop down all stems to within 60cm (24in) of the ground. Train new growth by tying in the best stems and removing wayward shoots.

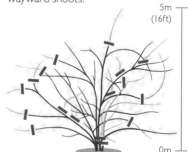

5m (16ft)

0m

Shorten flowered stems and remove dead, crossing, and congested shoots.

KERRIA _KERRIA_

This medium-sized, deciduous shrub is hardy, fast-growing, and easy to care for. In fact, kerria is the ultimate grow-anywhere plant with interest throughout the year: from its golden-yellow, cup-shaped spring flowers to fine summer and autumn foliage, then bright green stems through winter.

PLANT TYPE Deciduous shrubs
HEIGHT Up to 3m (10ft)
SPREAD Up to 3m (10ft)
FLOWERS ON Last year's stems, in spring
LEAF ARRANGEMENT Alternate
WHEN TO PRUNE After flowering
RENOVATION Yes

After flowering, kerria requires annual pruning to keep its vigorous, long, arching stems in check.

The white-marked leaves on _Kerria japonica_ 'Picta' enhance its blooms.

HOW THEY GROW

Kerria is a tough and adaptable shrub that can be used as a specimen plant or massed in a woodland setting. It is relatively deer-resistant and is also good for planting on slopes or anywhere soil erosion is a problem. It has an upright habit with arching stems, which gives it a graceful character.

However, it is extremely vigorous in certain conditions, and it spreads readily through suckers, quickly developing a thicket of stems. In some regions it is seen as an invasive plant.

It grows in moist but well-drained soil and is an excellent choice for a difficult shady corner, even a north-facing one. It also grows in sun if the soil is moist enough, but try to avoid direct afternoon sun to prevent the flowers from bleaching and becoming paler.

Kerria japonica 'Pleniflora' is a popular variety with double flowers like mini-chrysanthemums, while the less vigorous form _K.j._ 'Picta' offers the bonus of variegated, cream-edged leaves.

HOW TO PRUNE

Kerria are fast and strong growing, so prune annually to keep them in check and maintain vigour. After flowering, remove dead wood and thin out older stems — some gardeners prune out one in three stems each year. Shorten flowered stems back to strong sideshoots or, to keep to size, cut them down to buds lower down on the branches. To control unwanted spread from the creeping roots, remove suckers in autumn or winter. Follow the growth back to the base of the shrub, scraping off soil if necessary, and tear it off the parent plant — never cut it off.

RENOVATION PRUNING Kerria usually responds well to hard pruning. To rejuvenate an overgrown shrub, cut it all down to 30cm (12in) above the base — or even almost to ground level.

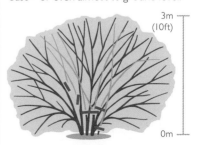

3m (10ft)

0m

Thin out old stems to ease congestion, and remove suckers at the base.

LABURNUM *LABURNUM*

Laburnums are small trees that are often trained as standards, or over arches, pergolas, and even tunnels. They are grown for their bright yellow, pea-like blooms, which make a jaw-dropping spectacle, like golden rain, hanging down in long racemes in late spring and early summer.

PLANT TYPE Deciduous trees
HEIGHT Up to 8m (26ft)
SPREAD Up to 8m (26ft)
FLOWERS ON Last year's stems, in late spring–early summer
LEAF ARRANGEMENT Alternate
WHEN TO PRUNE Late summer–early winter
RENOVATION No

HOW THEY GROW

When left to their own devices, laburnums are upright trees with an open spreading habit. Although they are typically grown as freestanding specimens that need little attention, they are also trained as standards or over structures.

The grey-green leaves are grouped in threes and have downy undersides. They accompany the profuse racemes of golden flowers in summer followed by the large seed pods, which change from light green to brown or black as they ripen later in the year.

> **TOP TIP** ALL PARTS OF LABURNUM ARE POISONOUS SO NEVER INGEST, AND WEAR PROTECTION SUCH AS GLOVES WHILE HANDLING THIS TREE. TRY TO PREVENT PETS AND YOUNG CHILDREN FROM REACHING THE FLOWERS OR SEED PODS.

Laburnums prefer light soils and full sun, but will grow in any aspect in any moderately fertile, well-drained soil; they hate to be waterlogged, however. They are fully hardy, and vigorous.

Common laburnum (*L. anagyroides*) has some interesting cultivars including *L.a.* 'Aureum', which has yellow leaves, and *L.a.* 'Pendulum', which has a weeping habit that makes it an excellent choice for a standard tree.

Laburnum × *watereri* 'Vossii' carries very long flower racemes, up to 60cm (24in). As with many laburnums, this cultivar is commonly available as a grafted plant, so it may produce suckers from the rootstock. Remove their buds or shoots, which develop below the graft union, as soon as you see them.

HOW TO TRAIN

Laburnums work well as central-leader standard trees, trained over several years from a young feathered tree, which has a single main stem off which grow a spread of lateral branches ("feathers"). When planting, secure the main stem to a strong cane, tree

All parts of laburnum are highly toxic so take extra care with this plant.

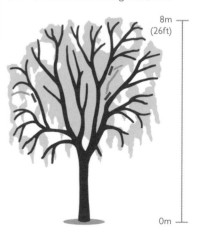
8m (26ft)

0m

Prune freestanding trees to remove only damaged or awkward growth.

Laburnum × *watereri* **'Vossii'** bears its trailing flowers in early summer.

CREATING A
LABURNUM TUNNEL

Training a laburnum over an arch or series of arches as a tunnel is not difficult, but does require ongoing regular pruning to form and maintain the feature. The first consideration is to construct a permanent solid support in a strong, long-lasting material like metal, which will not rot or require upkeep. Then source young plants, preferably feathered trees (see *How to train, left*), that are less than two years old, because they will be more flexible and easier to train. In spring, plant them on either side of the arch, or, in the case of a tunnel, in a row along each side, set 2–3m (6½–10ft) apart. Tie the central stem of each plant to the framework of the arches, and then secure strong lateral branches along the sides. If there is not enough horizontal growth to begin with, you may have to shorten the central stem to encourage the plant to produce more.

As this framework develops and the plants establish, continue tying in growth and pruning each year in early winter. Shorten any dead wood, cutting to just above where the dead bit meets the live growth. Remove unwanted new shoots, and shorten the whippy growth on those that remain to within two or three buds of the main framework. This will encourage spur production, which will result in more flowers.

Create your own show-stopping tunnel like the famous Laburnum Walk at Barnsley House in Gloucestershire, UK.

TOP TIP IT CAN BE HARD TO WORK ACROSS THE TOP OF AN ARCH OR TUNNEL, SO TO MAKE IT EASIER TO ACCESS THE PLANTS BEING TRAINED ON SUCH A SUPPORT, GENTLY LIFT THEM OFF AND LAY THEM FLAT ON THE GROUND TO DO THE PRUNING. THEN SECURE THEM BACK ON THE TOP OF THE STRUCTURE.

stake, or vertical support; loosen the tie as the stem expands. Start training in the first year after planting, in early winter. On the top-third of the plant, only take out dead, damaged, or crossing growth. Reduce the lateral branches of the central-third by at least one-half, and remove all laterals on the bottom-third. Repeat in the second and third years, to develop a trunk clear to the desired height, usually to around

1.5m (5ft). If you try to remove branches in subsequent years it may leave large holes in the main stem.

HOW TO PRUNE

Freestanding specimen trees are best left alone. They do not need regular attention, except to remove dead or damaged stems and any badly placed or crossing growth. If this is necessary, you

should prune between late summer and early winter, when the plant is dormant; in any other season it will bleed sap.

Laburnums that have been trained as standards or over a permanent support should be worked on in early winter, as detailed above.

It is best to replace unproductive or old laburnum trees as they do not respond well to renovation and, in any case, they can be short-lived.

LAGERSTROEMIA *LAGERSTROEMIA*

Few plants can compete with lagerstroemia for flounce and flamboyance, with its masses of crinkly frilly flowers stealing the show in summer. They offer more than pretty blooms, however, by producing wonderful autumn leaf colour and, in mature specimens, interesting peeling bark, too.

PLANT TYPE Deciduous and evergreen trees and shrubs
HEIGHT Up to 8m (26ft)
SPREAD Up to 8m (26ft)
FLOWERS ON Current year's stems, in summer–autumn
LEAF ARRANGEMENT Opposite
WHEN TO PRUNE Early spring
RENOVATION Yes

***Lagerstroemia indica* 'Seminole'** carries "crepe-paper" blossom in summer.

HOW THEY GROW

Mature lagerstroemias are typically small trees, either single or multi-stemmed, which form an open vase shape and have an upright spreading growth habit. However, young plants and those grown in cool-temperate climates are more likely to grow as shrubs, with a bushy habit.

Lagerstroemia is relatively slow-growing. As it is drought-tolerant, you can grow it in any well-drained soil, except clay. Its flowers are long-lasting, and the white, pink, purple, or red blooms are loved by insects. They appear in cone-shaped panicles of ruffled petals, reminiscent of crumpled tissue paper. The plant needs a hot summer to produce a good display. The leaves develop shades of yellow, red, and orange in autumn; and in winter the grey and brown trunk peels to reveal fresh pink to cinnamon-coloured bark.

Many lagerstroemia species are from tropical regions and are not frost hardy, but there are some suitable for warm-temperate zones, such as crepe myrtle (*Lagerstroemia indica*). It tolerates temperatures down to −5°C (23°F) if planted in a sheltered position, such as by a south- or west-facing wall, in full sun. Alternatively, in colder areas, crepe myrtle can be grown in a container that is brought into a conservatory or greenhouse for winter.

HOW TO TRAIN

Lagerstroemia can be trained to your desired shape. To encourage a multi-stem effect that shows off the bark in a shrubby specimen, choose an odd number of strong upright stems, and cut all others back to the ground. Remove the lower branches on the remaining stems.

HOW TO PRUNE

Established plants do not require much pruning beyond taking out dead, diseased, and damaged wood, as well as any wayward or crossing shoots, in early spring. To encourage more flowers, you can also thin out the crown each year to allow air and light into the centre of the plant, removing congested and inward-growing shoots, and pruning any weak growth less than the diameter of a pencil back to the nearest shoot or stem.

RENOVATION PRUNING Lagerstroemia responds to hard pruning and renovation, producing fresh new growth. Some gardeners cut their lagerstroemia right down to a low framework about 60cm (24in) from the base every year. However, this encourages weak growth with lots of foliage and fewer later flowers, and means you miss out on the beautiful bark that develops on mature specimens, so it should be done only if necessary to revive or reshape an overgrown plant.

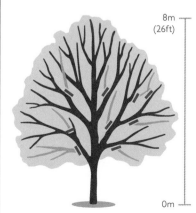

8m (26ft)

0m

Remove crossing, inward-growing, and weak shoots back to the nearest branch.

BAY *LAURUS*

Bay is grown for its dark green, leathery foliage, which stays on the plant year-round. The leaves are aromatic and edible, and often used for culinary purposes to enhance the flavour of a range of dishes. These trees and shrubs are also suitable for clipping into neat formal shapes.

PLANT TYPE Evergreen trees and shrubs
HEIGHT Up to 7.5m (25ft) unless clipped
SPREAD Up to 3m (10ft)
FLOWERS ON Last year's stems, in spring
LEAF ARRANGEMENT Alternate
WHEN TO PRUNE Free-growing specimens in spring; topiary in summer
RENOVATION Partial

HOW THEY GROW

Sweet bay (*Laurus nobilis*) is a dense but slow-growing shrub or small tree with an upright habit and loosely conical shape. Though most appreciated for its evergreen qualities, it also has small, yellow, spring flowers, which are followed by black berries on plants that are female. It has smooth, ovate, green leaves, though *L.n.* 'Aurea' has golden foliage.

This tree or shrub, however, is often clipped into topiary shapes such as cones or is trained as a standard with a single straight stem or with several stems twisted together. As well as being popular for formal garden schemes, bay can be grown in planters used to flank a doorway or to line a path.

The cheerful leaves on *Laurus nobilis* 'Aurea' brighten up the winter garden.

Bay is often clipped into standard topiary shapes such as this "lollipop" one.

It grows in fertile, moist but well-drained soil in full sun or partial shade. Mature specimens are usually hardy down to around −5°C (23°F), but young plants are more tender. Plant bay in a sheltered spot, perhaps against a warm wall, to protect it from frost and cold drying winds in cool-temperate climates. Bay is less hardy if growing in a container, so overwinter it in a frost-free space such as a conservatory or greenhouse.

HOW TO PRUNE

FREE-GROWING TYPES Bay needs minimal pruning but, with a regular trim, will retain its dense habit and be easier to keep in shape and to its allotted space. In spring, remove any dead and diseased shoots as well as any leaves or shoots damaged by frost.

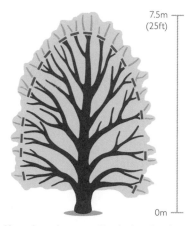

7.5m (25ft)

0m

Keep bay dense and bushy by trimming and by removing frost-damaged shoots.

Take out any weak, wayward, crossing stems, any that rub, and inward-growing shoots, and cut any overlong ones back to healthy leaf nodes.

TOPIARY Prune clipped shapes regularly to encourage dense growth and to maintain their form. Do this in summer, at least twice. Trim shoots with a pair of sharp secateurs, clipping back to a bud facing the direction you wish each shoot to grow.

RENOVATION PRUNING Large established bays in need of rejuvenation respond to being cut back hard, but regrowth can take a long time as they are slow-growing. As a result, it is best to renovate in stages over two or three years. Each spring, reduce the plant by at least one-third.

LAVENDER *LAVANDULA*

This Mediterranean plant loves hot sunny conditions and offers year-round, grey-green foliage interest. Its crowning glory, however, are its spikes of intensely fragrant, purple flowers, which are often picked and dried for scent, or used in the kitchen to season sweet and savoury dishes.

PLANT TYPE Evergreen shrubs and subshrubs
HEIGHT Up to 1m (3ft)
SPREAD Up to 1.2m (4ft)
FLOWERS ON Current year's stems, in late spring–late summer
LEAF ARRANGEMENT Opposite
WHEN TO PRUNE After flowering
RENOVATION No

HOW THEY GROW

This easy-care, rounded or spreading bushy shrub bears evergreen foliage in silver or grey-green, and long-stalked flowers in shades of blue, violet, and purple but also occasionally pink and white. The heady scented blooms are a magnet for bees and other pollinators.

There are several different types of lavender, and they vary in hardiness. Hardy species include English lavender (*L. angustifolia*) and its popular cultivars *L.a.* 'Munstead' and *L.a.* Hidcote Group as well as taller-growing lavandin (*L. × intermedia*) and its varieties such as *L. × i.* 'Grosso'.

More borderline types that are frost hardy to about −5°C (23°F) include French lavender (*L. stoechas*), also called Spanish or butterfly lavender, which bears petal "ears" at the top of each flower. Half-hardy species such as *L. dentata* need temperatures above 0°C (32°F). All borderline types require a sheltered position or, in cool-temperate climates, to be grown in containers, which can be brought indoors over winter.

Being drought-tolerant, lavenders hate wet conditions so plant in well-drained soil that is neutral to alkaline, in full sun. Those grown on lighter soils live longer than those on heavy soil.

HOW TO PRUNE

Prune lavender every year to prevent it becoming gappy and leggy at the base. It does not regenerate from bare wood, so, once overgrown, it will have to be replaced rather than renovated. Trimmed regularly from the beginning, however, it will stay dense and floriferous for years – in the right conditions, plants have been known to thrive for two decades or more.

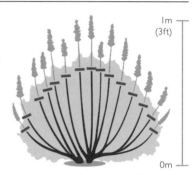

Trim at least one-third off flowered stems when flowering has finally finished.

HARDY TYPES Once the flowers have finished blooming, shorten the flowered stems with sharp secateurs by at least one-third to just above where fresh little buds are visible on the stem, or on established plants remove up to two-thirds of each flowered stem. As long as you cut above where the lowest, three or four, new buds are developing on each stem, you can chop right down to about 23cm (9in) from the base if necessary. Some gardeners also like to give English lavender another light trim in spring, with shears, to help keep a neat hummock shape, although this delays flowering.

MORE BORDERLINE TYPES Deadhead as flowers fade in late spring, to encourage repeat-flowering during summer. Then trim lightly all over after flowering has finished completely, being careful to leave at least 2.5cm (1in) of the current year's growth and making sure that you do not cut into old wood.

English lavender cultivars 'Munstead' and Hidcote Group are highly fragrant.

You may get a second flush of blooms if you cut off the flowered stems.

TREE MALLOW *LAVATERA*

Tree mallows are cultivated for their large, pink or sometimes white, hollyhock-like flowers, which are produced over a long period in summer. Not only is this the perfect plant for the back of a mixed border, but it is also easy to care for and fast-growing, if quite short-lived.

PLANT TYPE Semi-evergreen and deciduous shrubs and subshrubs
HEIGHT Up to 2m (6½ft)
SPREAD Up to 1.2m (4ft)
FLOWERS ON Current year's stems, in summer–autumn
LEAF ARRANGEMENT Alternate
WHEN TO PRUNE Spring
RENOVATION Yes

Lavatera × clementii **'Barnsley'** has felted leaves and soft pink flowers.

Very vigorous *Lavatera × clementii* 'Rosea' is a long-flowering tree mallow.

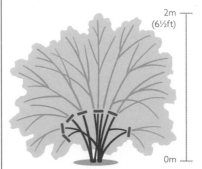

Cut back stems when new growth starts to appear in spring.

HOW THEY GROW

Tree mallow has an upright bushy habit with slightly arching stems and deeply lobed, green foliage. The big, saucer- or funnel-shaped blooms appear in midsummer and continue right up to the first frost; they range in colour from white to dark pink. The shrubby types, including *L. × clementii* 'Barnsley' and *L. × c.* 'Rosea', are mostly hardy but can be damaged by severe frosts and cold winds, so in cool-temperate climates plant them in a warm sheltered position such as by a south- or west-facing wall. They are, however, quite salt-tolerant so a good choice for coastal gardens.

Tree mallows grow very fast. Fresh growth can be floppy and in need of staking, while older woody growth can become brittle, and is easily damaged by wind. They prefer poor, dry, free-draining conditions but do tolerate most soils as long as they are well-drained. They flower better in full sun than partial shade.

HOW TO PRUNE

These plants will last longer, produce more flowers, and be less likely to need staking if they are pruned regularly, which will encourage fresh stems to emerge from the base. Cut back in spring, when new growth emerges on the plant, and as soon as the risk of hard frost has passed. Initially remove old woody stems completely, along with any weak growth. Then reduce the strong-growing stems that remain to a framework about 30cm (12in) from the base.

In summer and autumn, look out for shoots damaged by wind and cut back to the nearest healthy bud.

RENOVATION PRUNING Tree mallow responds quickly to drastic pruning, so to renovate you can cut back all stems hard, almost to the base, in spring as fresh growth begins.

2m (6½ft)

0m

Reduce all vigorous stems to a framework close to the base each year.

LEYCESTERIA *LEYCESTERIA*

Leycesteria are vigorous, low-maintenance shrubs with a long season of interest, from their pendent racemes of tubular flowers in summer to their fleshy autumn fruits. The nectar-rich blooms attract bees while the berries are loved by birds, making these bushes perfect for wildlife.

PLANT TYPE Deciduous shrubs
HEIGHT Up to 2m (6½ft)
SPREAD 2m (6½ft) or more
FLOWERS ON Current year's stems, in late spring–early summer or summer–autumn
LEAF ARRANGEMENT Opposite
WHEN TO PRUNE After flowering or in early spring
RENOVATION Yes

Himalayan honeysuckle has upright stems that arch out with dangling flowers.

HOW THEY GROW

These suckering shrubs have an upright bushy or arching habit. They produce long pointed leaves and distinctive, shrimp-like, hanging flowers, which develop into edible berries that are said to have a toffee flavour. They make good specimen shrubs and thrive in woodland gardens and coastal areas. Plant in any moist but well-drained soil, in full sun to partial shade.

Himalayan honeysuckle (*L. formosa*) is extremely tough and usually fully hardy. Plant it in a sheltered position for protection in regions that regularly

Maroon bracts shield the flowers, then berries, on this hardy shrub.

experience hard frosts. Even when damaged by frost, however, it will bounce back quickly. Its white flowers, framed by maroon bracts, appear over a long period from summer to autumn, and are followed by purple-red berries. Its cane-like, hollow stems are greenish blue initially and will rapidly form a thicket-like clump. *Leycesteria formosa* GOLDEN LANTERNS has yellow foliage.

Leycesteria crocothyrsos is more tender but will also rebound quickly from the base after a cold winter. It has an arching habit, veined leaves, and yellow flowers in late spring and early summer, followed by green berries.

HOW TO PRUNE

Leycesteria can be managed with minimal pruning, simply trimming back the flowered shoots after flowering. However, it is a good idea to thin out clumps by removing weak growth and cutting back thick old stems to the ground. This will prevent congestion at the centre of the plant and stems going bare towards the base, and will encourage fresh stems. Some gardeners choose to cut the whole plant back to a low permanent framework around 30cm (12in) from the base in early spring each year, to encourage vigorous fresh growth.

RENOVATION PRUNING Leycesteria is so resilient that, if it is old and overgrown or in need of reviving, it can be cut back hard right to ground level in early spring, and it will respond quickly. It is a good idea to do this every few years to stop the plant from getting straggly.

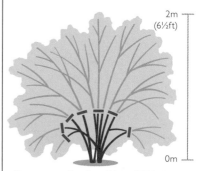

2m (6½ft)

0m

Cut stems back to 30cm (12in) above the ground in early spring.

PRIVET *LIGUSTRUM*

This robust, hardy, long-lasting shrub is most commonly cultivated for its small, soft, smooth-edged, green leaves and dense habit. It provides shelter, summer flowers, and berries for birds and pollinators. When grown as a hedge, it quickly creates privacy at garden boundaries, too.

PLANT TYPE Evergreen shrubs
HEIGHT Up to 4m (13ft)
SPREAD Up to 4m (13ft)
FLOWERS ON Current year's stems, in summer
LEAF ARRANGEMENT Opposite
WHEN TO PRUNE Free-growing specimens after flowering; hedges in summer
RENOVATION Yes

HOW THEY GROW

Privet has an upright or bushy habit, which makes it perfect for hedging, yet develops an open vase shape when free growing. In harsh winters, it may shed some of its leaves, but it recovers quickly in spring.

The two most common types are common or wild privet (*L vulgare*) and vigorous garden privet (*L. ovalifolium*). Garden privet clips more neatly, so is good for formal hedging and topiary, but wild privet offers more benefits for wildlife. When left unclipped, both produce white summer flowers, borne in panicles, which are beloved by bees and other insects. They are followed by dark purple or black berries, which are poisonous, so do not plant privet at a boundary where livestock can access it.

Privet is tough, and easy to grow. Being tolerant of air pollution, it is popular for urban gardens, and can also manage the salt spray of coastal areas.

However, street-front plants may be damaged by concentrated splashback from ice-preventing road salt in winter.

Privet grows in full sun or partial shade in any moist but well-drained soil except for boggy clay. Variegated types such as *L. ovalifolium* 'Argenteum' (with cream-edged leaves) and *L.o.* 'Aureum' (with yellow foliage) have better colour when planted in sun.

HOW TO TRAIN

To create a hedge, in the first few years after planting, gently trim in summer to encourage bushiness.

HOW TO PRUNE

FREE-GROWING TYPES Trim privet shrubs after flowering by shortening flowered shoots and removing any dead, damaged, and crossing stems, as well as thinning out congested old stems that have become bare at the base.

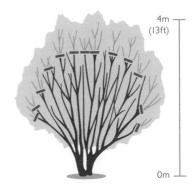

4m (13ft)

0m

Trim back flowered shoots after flowering on free-growing shrubs.

HEDGES Once the hedge is established in an appropriate shape, clip it in summer, after birds have flown their nests, using sharp hedging shears or an electric hedge trimmer. For a light neat form, clip twice: once at the beginning of summer, and again at the very end.

RENOVATION PRUNING Privet regenerates from old wood, so if a free-growing specimen or hedge is overgrown it can be pruned back to your desired size provided it produces plain green leaves. Never renovate a variegated privet, because it could cause the leaves to revert to plain green. To retain coverage and to give each plant its best chance of recovery, cut it back while dormant, in stages. At the end of winter in the first year, reduce the top and one side. If the plant has produced good regrowth, you can cut back the other side the following year; if not, wait another year before doing this.

Garden privet produces bee-friendly, white flowers when left unclipped.

To keep a privet hedge in shape, clip with shears once or twice in summer.

HONEYSUCKLE *LONICERA*

Honeysuckle is best known as a wonderful, wildlife-friendly climber with nectar-rich, scented flowers in shades of white, cream, pink, yellow, and orange, but there are also shrubby types with varied growth habits, which are used for hedging and topiary as well as specimens in a border.

PLANT TYPE Deciduous and evergreen climbers and shrubs
HEIGHT Up to 10m (33ft)
SPREAD Up to 2.5m (8ft)
FLOWERS ON Current or previous year's stems, in spring, summer, autumn, or winter
LEAF ARRANGEMENT Opposite
WHEN TO PRUNE After flowering for most types; evergreen climbers in spring; hedges in spring, summer, and autumn if necessary
RENOVATION Yes

Common honeysuckle has nectar-rich blooms and is the best choice of honeysuckle for wildlife.

Lonicera x *purpusii* **'Winter Beauty'** bears its dainty fragrant flowers in winter.

HOW THEY GROW

A diverse range of honeysuckles is available, including climbing and shrubby, evergreen and deciduous, as well as early- and late-flowering types.

CLIMBERS The climbing honeysuckles have flexible twining stems, which wrap around any support. Evergreen types will quickly clothe a wall or fence and give green foliage cover year-round. Deciduous types put on more of a floral show in spring or summer with their whorls of trumpet-shaped blooms, often deliciously scented.

They do best in fertile, moist but well-drained soil in dappled shade, though they tolerate full sun. They even grow in full shade but will not produce flowers.

Evergreen *L. japonica* 'Halliana' is a hardy vigorous climber that, from spring through to late summer, produces its heavily fragrant, white blooms fading to cream, and later black berries.

Deciduous common honeysuckle (*L. periclymenum*), also known as woodbine, with its young red stems and oval, grey-green leaves, is a great choice for growing up a wall or fence or for scrambling through other shrubs, and

it is really excellent in a wildlife garden. The summer flowers are magnets for pollinators, and the red berries that follow, though not suitable for human consumption, are a real treat for birds and other wildlife. *Lonicera periclymenum* 'Belgica', with its white and red flowers, works better than the species in a garden setting.

SHRUBS The shrubby types of honeysuckle vary in habit from low compact plants, which work as ground cover, to those with upright stems and arching spreading growth. When left unclipped, they develop a loose and lax habit, sometimes having an unkempt look.

TOP TIP TO ATTRACT POLLINATING MOTHS, SUMMER-FLOWERING HONEYSUCKLES HAVE A STRONGER SCENT AT NIGHT TIME THAN DURING THE DAY. THEREFORE, PLANT THEM NEAR AN AREA WHERE YOU LIKE TO SPEND TIME IN THE EVENING SO YOU CAN ENJOY THE NOCTURNAL FRAGRANCE AND THE WILDLIFE.

Deciduous shrubs are usually grown for their flowers, which in many types appear in winter. *Lonicera* × *purpusii* 'Winter Beauty' is a deciduous rounded shrub that grows to 2m (6½ft) by 2.5m (8ft), and produces highly scented, white and yellow flowers in winter or early spring. Another good choice of deciduous honeysuckle for winter flowers is *L. fragrantissima*.

Evergreen types are commonly used as hedging. *Lonicera ligustrina* var. *yunnanensis* is a fast-growing, bushy shrub to 3m (10ft), with very small, dark green leaves, and is increasingly being used as a replacement for box (*Buxus*) hedging and topiary. *Lonicera ligustrina* var. *yunnanensis* 'Baggesen's Gold' bears bright yellow foliage.

Shrubby honeysuckles grow in well-drained soil in full sun or partial shade, but they dislike extremes such as heavy soil or dry conditions, and winter-flowering types should be sheltered from winds.

HOW TO PRUNE

Regardless of the type of honeysuckle, when pruning always cut out dead, damaged, and diseased stems, inward-growing and crossing shoots, and any overlong or wayward stems. Remove some older stems back at the base, to ease congestion.

CLIMBERS For evergreen, usually late-flowering types, which flower on the current season's growth, prune in spring, as already described (see *left*). Also give a very light trim, pruning stems to strong buds lower down, and thin some stems at the top of the plant.

For deciduous, usually early-flowering types, which produce their blooms on shoots from the previous year's growth, prune after flowering, as already described (see *left*). Also shorten flowered shoots back to healthy buds.

SHRUBS For evergreen and deciduous, summer- and winter-flowering shrubby honeysuckles, prune minimally after flowering, as already described (see *left*). Shrubby evergreens such as *L. ligustrina* var. *yunnanensis*, when grown freely as bushes, may require a trim in spring and in summer, to keep them neat.

HEDGES These require more than one trim to keep them in shape. Clip them in spring and midsummer; also in early autumn if necessary.

RENOVATION PRUNING Without regular attention, climbing honeysuckles can become a tangled mess, with bare

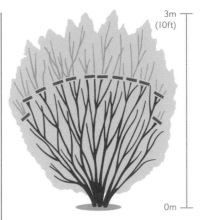

Free-growing evergreen shrubs can be trimmed more than once a year.

stems towards the base. To rectify this, shorten all stems down to 60cm (24in) from the base, in winter, and cut back the oldest and weakest stems completely. The plant will respond quickly with new growth, but probably won't flower that year. Overgrown honeysuckle bushes and hedges can also be cut back hard in this way, in late winter or early spring, but will benefit from a staged renovation, taking out one in three of the oldest stems each year for three years.

TOP TIP WHEN TAKING OUT WHOLE STEMS ON CLIMBING HONEYSUCKLE, IT IS BEST TO CUT AND REMOVE THEM IN SECTIONS. IT MIGHT ALSO BE EASIER TO DO THIS ON DECIDUOUS TYPES WHEN THEY LOSE THEIR LEAVES IN WINTER.

Cutting back dead growth from a climbing honeysuckle is easier with shears.

Prune deciduous, winter-flowering types after flowering, with secateurs.

Renovate an old climber by cutting stems down to 60cm (2ft) from the base.

MAHONIA *MAHONIA*

These architectural evergreen shrubs have spiny, glossy green leaves and scented yellow flowers, followed by purple-blue to black berries. As well as offering colour during the winter months, the bright blooms and fruit are invaluable for early foraging insects and birds.

PLANT TYPE Evergreen shrubs
HEIGHT Up to 4m (13ft)
SPREAD Up to 4m (13ft)
FLOWERS ON Last year's stems, in late autumn, winter, and spring
LEAF ARRANGEMENT Alternate
WHEN TO PRUNE After flowering
RENOVATION Yes

HOW THEY GROW

Mahonias are tough and hardy plants with diverse growth habits, from low and spreading, to open and suckering, and to tall and upright. *Mahonia repens*, which grows to just 30cm (12in), can be used as ground cover, while others are more suitable for the back of a border or a woodland garden. There are different forms of flower, too, from the clustered panicles of *M. aquifolium* 'Apollo' (in spring) and *M. fortunei* (in autumn) to the upright, spreading, lemon-yellow spikes of *M. × media* 'Charity' (from late autumn to late winter) with its flowers over 30cm

(12in) long. *Mahonia eurybracteata* subsp. *ganpinensis* 'Soft Caress' is another interesting variety, with spine-free foliage, spire-like, honey-scented, yellow blooms in late summer to autumn, and a compact habit to 1.2m (4ft) tall.

Mahonias thrive in moist but well-drained soil that is moderately fertile. They prefer full or partial shade but will grow in full sun if the soil is kept moist. Shelter them from cold winter winds, which can damage the leaves.

HOW TO PRUNE

Mahonias require only minimal pruning and should be left to establish before beginning an annual regimen. After flowering, remove dead and damaged wood as well as crossing growth, and trim back any shoots that ruin the shape of the plant.

Ground-cover specimens can be trimmed each year, and cut back to near the base every few years to keep them vigorous. For the smaller-growing, usually spring-flowering types, which are typically mound-forming or spreading, encourage fresh growth and help light and air get to the centre of the shrub by pruning up to one-quarter of older stems at a healthy bud near the base. Cut out suckers where they appear, and remove outer stems to keep shrubs to size.

RENOVATION PRUNING The taller, upright-growing mahonias, which are typically winter-flowering, can become

The long leaves and floral sprays make *Mahonia × media* an impressive sight.

Shorten bare stems, to encourage new ones to develop farther down.

bare at the base over time and they develop a tree-like form. If you would prefer to have leafy growth lower down the plant, cut stems back hard, to 15–30cm (6–12in) above the ground, immediately after flowering in late winter to early spring. Renovation pruning can be done all at once or be staggered over a few years.

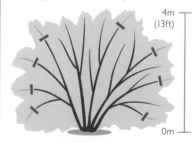

After flowering, trim back any shoots that spoil the plant's symmetry.

CRAB APPLE *MALUS*

Crab or ornamental apple trees are wonderful garden specimens, producing beautiful blossom in spring followed by fruits of various shapes, sizes, and colours, some of which can be used to make preserves. Many hold their fruit until late in the season and also have colourful autumn foliage.

PLANT TYPE Deciduous trees
HEIGHT Up to 12m (39ft)
SPREAD Up to 5m (16ft)
FLOWERS ON Previous year's stems, in spring
LEAF ARRANGEMENT Alternate
WHEN TO PRUNE Late winter
RENOVATION Partial

HOW THEY GROW

Most of these trees have an upright growth habit, though there are weeping types available. They are hardy, easy-to-grow trees offering almost year-round interest despite being deciduous, from their bee-friendly, white or pink blossom to the burnished autumn leaves and colourful fruit clinging to winter branches.

Crab apples are usually small to medium-sized trees, with some such as M. 'Evereste', at 7m (23ft) tall, remaining a suitable size for even small gardens. *Malus hupehensis* tends to grow much larger, to 12m (39ft), and has some of

Malus × robusta 'Red Sentinel' retains its red fruits right into winter.

the best autumn leaf colouring. Other popular types include M. × *robusta* 'Red Sentinel', with its white flowers followed by dark red fruits, and M. × *zumi* 'Golden Hornet', with its pink-budded, white flowers and big yellow fruits that are perfect for making crab apple jelly.

Check the eventual height and spread of cultivars before you buy any crab apple tree, to make sure it is suitable for the space available. Plant in moist but well-drained soil in full sun.

HOW TO PRUNE

After planting, crab apples need little pruning and should be allowed to develop their natural form. To keep the tree healthy, however, in late winter or early spring remove any dead, damaged, and diseased branches back to their base. Also cut out any suckers growing

Malus 'Butterball' combines its golden autumn leaves with golden fruits.

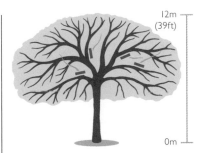

12m (39ft)

0m

Remove any crossing and inward-facing stems on crab apple trees.

from the ground or beneath the graft point on the tree, as well as any shoots that appear on the trunk or older limbs, often around old pruning wounds. If necessary, prune back any crossing or rubbing shoots, and any inward-facing growth, to keep a well-spaced canopy with good air circulation. If access is required under the tree for passage or mowing, consider raising the crown by removing the lowest limbs on the trunk (see p.31 for details).

RENOVATION PRUNING It is generally inadvisable to cut back a crab apple tree hard or to try a drastic renovation. However, if you have inherited an overgrown crab apple, you can attempt to restore it by following the pruning process described above. At the same time, reduce the length of overlong branches to fit in better with the overall shape of the tree by cutting them back to a side branch that is strong and at least one-third the diameter of the branch you are removing.

OLEANDER *NERIUM*

This tough bushy shrub may grow to form a small tree. It has narrow, grey-green, leathery leaves and white, pink, or red, sometimes scented, funnel-shaped flowers in summer. Oleander is drought-tolerant and suits a Mediterranean climate. Take care, though, as it is highly toxic.

PLANT TYPE Evergreen small trees and shrubs
HEIGHT Up to 2.5m (8ft)
SPREAD Up to 1.5m (5ft)
FLOWERS ON Current year's stems, in summer–autumn
LEAF ARRANGEMENT Opposite
WHEN TO PRUNE Late winter; pot-grown plants in late autumn
RENOVATION Yes

HOW THEY GROW

Oleander (*Nerium oleander*) has an upright growth habit and develops a rounded form. It is usually grown as a freestanding shrub or hedge outdoors in warm-temperate and dry-subtropical climates. In cool-temperate gardens, it is frost hardy to −5°C (23°F) in mild and coastal areas, but otherwise is best grown in a container and, before winter, moved into a conservatory or greenhouse – somewhere under glass where the temperature will stay around 10°C (50°F). It can be brought back outside each summer. This shrub prefers full sun but will grow in any moist but well-drained soil, and it

Nerium oleander 'Casablanca' blooms reliably through summer.

appreciates a sheltered, south-facing aspect. Oleander thrives in heat and, once established, is drought-tolerant in the ground. When grown in pots, water when dry during the growing season but only sparingly when indoors overwinter. Repot containerized oleanders regularly.

The tubular blooms are borne over a long period in clusters at the ends of the stems, and are followed by long seed casings that split. Pot-grown oleanders will not flower or grow as well as those in the ground, and may not live as long. It is best to replace failing specimens.

HOW TO PRUNE

Every part of the plant is extremely poisonous, and the leaves and sap can irritate the skin. Therefore, always wear gloves and protective equipment when pruning oleanders to protect yourself. Even the oleander clippings should be composted (for ornamental garden use only) as opposed to burned, as the smoke is also toxic.

These plants require only minimal pruning, and can be allowed to develop into large specimens if grown in the ground outdoors in warmer areas or in beds under glass. Young plants can be tip-pruned to encourage them to develop a bushy form. Established plants may benefit from some shaping, so trim back any shoots that spoil the shape of the plant in late winter, and take out weak or crowded growth.

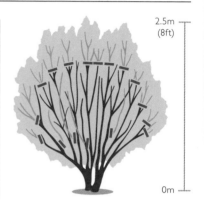

2.5m (8ft)

0m

Thin out congested stems and tip-prune to encourage bushiness.

CONTAINER-GROWN PLANTS If oleander is being grown in a container that has to be brought indoors overwinter, it will probably be necessary to keep it compact. In late autumn, therefore, you can clip it to size, reducing the flowered shoots by one-half and sideshoots back to two or three sets of leaves.

RENOVATION PRUNING Oleanders tolerate hard pruning, but can develop long, leafy, new growth in response to such drastic treatment, so it is best to renovate an overgrown specimen over the course of a few years.

In late winter, cut out one-third of the oldest stems to the ground. The following year, remove another one-third of old stems, and trim back the new shoots. In the third year, prune out the final one-third of old stems to near the base, and shorten the long new shoots.

DAISY BUSH *OLEARIA*

Daisy bush hails from **New Zealand** and is an easy-care, evergreen shrub or small tree. It has leathery foliage and tightly packed, composite, daisy-like flowers, which are sometimes fragrant, in white, pink, or purple. It is deer- and pollution-resistant, and tolerant of salt spray and wind.

PLANT TYPE Evergreen small trees and shrubs
HEIGHT Up to 4m (13ft)
SPREAD Up to 4m (13ft)
FLOWERS ON Last year's and current year's stems, in late spring or summer
LEAF ARRANGEMENT Opposite; occasionally alternate
WHEN TO PRUNE Spring; hedges also in summer
RENOVATION Yes

Olearia × scilloniensis is an evergreen shrub that will thrive in coastal locations and mild areas of cool-temperate regions.

HOW TO PRUNE

Prune these plants to keep them in shape and to the required size. Remove dead, diseased, and damaged stems in spring, once new growth appears, and trim back any wayward stems. Cut off the flowers after they have faded.

For fully hardy types that need to be reduced in size, shorten the previous season's growth by one-third to one-half. Frost hardy types will benefit from a light trim all over, to keep them compact.

HEDGES As well as a spring prune, these may require a second cut in summer, to maintain a neat form.

RENOVATION PRUNING Overgrown freestanding daisy bushes and hedges can be pruned hard, over three years. In spring, cut back one-third of the stems, starting with the oldest ones. Repeat these cuts in the following two years until the plant is rejuvenated.

HOW THEY GROW

Many daisy bushes are tender but there are some hardy types that are suitable for growing in cool-temperate climates. In general, they like sun and should be offered the protection of a sheltered position such as beside a warm wall. They grow in chalk soils, but dislike clay as they prefer soil to be well-drained. Drought-tolerant daisy bush is a good choice for a border specimen or to help stabilize a slope.

One of the hardiest daisy bushes is *O. macrodonta*, a vigorous, medium-sized shrub with a bushy habit, which is frequently used for hedging. It is easy to grow, tolerating heavy soils and any aspect. It bears large clusters of small, scented, white flowers, which are attractive to bees and other pollinators. Extra value is offered by its pretty peeling bark.

Another hardy type, *O. haastii*, forms a dense bush with an upright growth habit, to 2m (6½ft) tall by 3m (10ft) wide. It has glossy green leaves with soft silver undersides, and in summer produces masses of white blooms. It can generally cope with temperatures down to −10°C (14°F).

Olearia × scilloniensis is rounder in form, with wavy-edged leaves that have soft silver undersides, and clusters of white daisy flowers. Its cultivar 'Master Michael' has a more spreading habit, and produces purple flowers with yellow centres in late spring. They need shelter from cold drying winds and are frost hardy down to about −5°C (23°F).

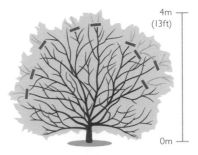

4m (13ft)

0m

Deadhead spent flowerheads once daisy bush has finished blooming.

OSMANTHUS *OSMANTHUS*

Serrated leaves and tubular, often fragrant flowers are the standout features of osmanthus, which is sometimes known as sweet olive or devilwood. It is often mistaken for holly (*Ilex*), but its leaves grow in pairs, rather than along the branch. It also has dark blue-black fruits.

PLANT TYPE Evergreen trees and shrubs
HEIGHT Up to 4m (13ft)
SPREAD Up to 4m (13ft)
FLOWERS ON Last year's and current year's stems, in spring, summer, autumn, or winter
LEAF ARRANGEMENT Opposite
WHEN TO PRUNE Late flowerers and *O. heterophyllus* hedges in spring; other hedge species, topiary, and early flowerers after flowering
RENOVATION Yes

Osmanthus × fortunei has leathery green leaves and highly scented flowers.

HOW THEY GROW

Osmanthus is an upright or dense, rounded, bushy evergreen shrub, occasionally with an arching or spreading growth habit. It is generally pest-free and typically very hardy, though most require protection from cold drying winds, so plant in a sheltered spot in the garden. Cultivate osmanthus in full sun or partial shade in any fertile, well-drained soil. Avoid growing it in waterlogged soil and overwatering this plant.

The small, funnel-shaped flowers are white or yellow, and often scented, like jasmine. They appear in clusters, mainly in spring or summer, though *O. armatus* blooms in autumn and *O. yunnanensis* in winter. *Osmanthus fragrans* f. *aurantiacus* carries orange blooms.

The toothed to spiny, year-round, dark green foliage makes species including *O. × burkwoodii*, *O. delavayi* and *O. heterophyllus* excellent choices for hedging and topiary. Osmanthus is also popular as a freestanding shrub for a border or woodland planting scheme. *Osmanthus heterophyllus* 'Goshiki' and *O.h.* 'Aureomarginatus' are holly-leaved, variegated forms.

HOW TO PRUNE

Only minimal pruning is needed for osmanthus, which is not particularly fast-growing. A regular light trim should keep the plant dense and compact.

EARLY-FLOWERING TYPES Every year, after flowering, prune osmanthus that bloom between winter and summer. Deadhead flowers, if the fruit is not required. Remove any dead, diseased, and damaged stems, and clip back overlong or wayward shoots.

LATE-FLOWERING TYPES In spring, before new growth begins, prune osmanthus that flower between

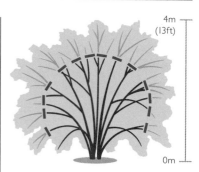

4m (13ft)

0m

Remove all faded flowers and clip back any particularly long shoots.

summer and autumn. If pruned any earlier, the resulting new shoots may suffer frost damage. As with early flowerers, remove spent blooms and dead or damaged stems, and trim any growth that ruins the plant's symmetry.

HEDGES AND TOPIARY Clip every year after flowering, and again, if necessary, a few months later. Trim *O. heterophyllus* hedges only in spring.

RENOVATION PRUNING Overgrown specimens can be renovated over a couple of years, cutting back in stages. In the first year, in spring, cut one-half of the oldest stems back to the base, and shorten the other stems by about one-half, to just above fat healthy buds. Also prune out stems and shoots that are congested, crossing, and rubbing. The following spring, remove the remaining oldest stems. It may take several years for flowering to resume after this drastic pruning.

VIRGINIA CREEPER

PARTHENOCISSUS

Being a very vigorous climber, Virginia creeper quickly clothes any surface with a mass of leafy growth, and then keeps going. It does produce small insignificant flowers, and some types also have berries, but it is generally grown for its autumn foliage in blazing hues of red and orange.

PLANT TYPE Deciduous climbers
HEIGHT Up to 30m (100ft)
SPREAD Up to 10m (33ft)
FLOWERS ON Last year's stems, in spring–early summer
LEAF ARRANGEMENT Alternate
WHEN TO PRUNE Late autumn–winter
RENOVATION Yes

HOW THEY GROW

Most of these deciduous climbers attach themselves to walls and fences with little adhesive pads (*see p.35*). Unusually, *P. vitacea* has twining tendrils so is more suitable for trailing from a support like a trellis, arch, or pergola. All Virginia creepers put on masses of growth each season and can also be used as ground cover; they spread far and wide, reaching a mounded height of about 30cm (12in).

Virginia creeper is well worth growing for its fresh green spring and summer foliage cover as well as the later colourful leaf display. Being deciduous, its large, lobed or hand-shaped leaves are shed by winter. The tiny, yellow-green blooms are barely noticeable when they appear in spring and early summer.

Other than keeping growth in check, these plants are easy care, and grow in any moist but well-drained soil, in full sun or partial shade, in any aspect. However, although they are fully hardy, in very exposed sites they may lose their autumn foliage earlier.

The best-known types include Virginia creeper (*P. quinquefolia*) and Boston ivy (*P. tricuspidata*). For a less vigorous one, try variegated *P. henryana*, which reaches 10m (33ft) tall.

HOW TO PRUNE

Most Virginia creepers need to be pruned on a regular basis to keep them under control and to a size suitable for most walls and other supporting structures. When grown against a house wall, they will happily romp into gutters and drains, between roof tiles, and over windows, and you should cut back these errant shoots as you see them, if practical. Time the yearly prune between autumn leaf fall and early winter. Cut back shoots hard to the allotted space, reducing growth to a woody framework of stems.

RENOVATION PRUNING Virginia creeper tolerates drastic pruning, and can be cut right back to about 60cm (24in) from its base. Do this any time between late autumn and late winter.

Boston ivy's green foliage turns an eye-catching russet-red in autumn.

The divided, five-lobed leaves of *Parthenocissus henryana* are silver veined.

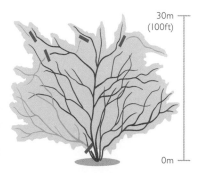

30m (100ft)

0m

Shorten overlong stems and prune out any bare unproductive ones.

PASSION FLOWER *PASSIFLORA*

Passion flowers are beloved for their extraordinary, eye-catching flowers, which have intricate forms with multiple parts including a central ring of colourful filaments – the "corona". These incredible blooms are followed, in some cases, by spherical or egg-shaped, yellow fruit that is edible.

PLANT TYPE Evergreen climbers
HEIGHT 3m (10ft) or more
SPREAD 3m (10ft) or more
FLOWERS ON Current year's stems, in summer–autumn
LEAF ARRANGEMENT Alternate
WHEN TO PRUNE Spring
RENOVATION Yes

Frost-hardy blue passion flower carries complex fragrant blooms.

The purple petals of *Passiflora* 'Amethyst' reflex backwards.

HOW THEY GROW

These vigorous woody climbers cling on to supports such as an arch or trellis with their winding tendrils. The leaves vary in shape, but are typically hand-shaped, dark green, and stay on the plant year-round. However, passion flowers are grown for their exquisite, showy, tropical-looking blooms, which are borne over a long period from summer to autumn. The fruits that follow on some types are edible, but not always tasty, and should not be eaten unless ripe (usually yellow). All other parts of the plant are toxic when ingested.

Most species of passion flower are native to tropical climates and are, therefore, tender and not suitable for growing outdoors in areas where the temperature drops below 7°C (45°F) in winter. However, some are frost hardy to fully hardy: for example, *P. incarnata* is fully hardy, yet still likes a sheltered position. The common blue passion

flower (*P. caerulea*), with its blue, green, and white flowers, is frost hardy to around −5°C (23°F), so can be grown outdoors in mild and coastal areas, but appreciates the protection of a warm, south-facing wall. *Passiflora caerulea* 'Constance Elliott' and *P.* 'Amethyst' are less hardy, but can make it through winter temperatures down to 0°C (32°F). Passion flowers will bloom more in warmer conditions. Offer extra winter protection such as horticultural fleece if frost is expected.

Grow in any moderately fertile, moist but well-drained soil, in full sun. If growing in a south-facing greenhouse or conservatory, offer shade from direct sun in the day, to protect the leaves from scorching.

HOW TO PRUNE

Passion flowers do not need regular pruning, but can be trimmed to fit their allotted space in spring. Remove any

dead, damaged, diseased, wayward, tangled, or bare and unproductive stems. Cut back frost-damaged stems and any overlong ones. After flowering, you can also neaten growth by shortening flowered shoots back to healthy buds, if the fruit is not required. Cutting back hard promotes leafy, non-flowering growth, so trim a little annually rather than a lot occasionally.

RENOVATION PRUNING Before replacing a passion flower that has become overgrown or bare at the base of stems, or that has suffered extensive frost damage, try to renovate it, even though such drastic treatment is not always successful. There will also be fewer flowers for a few seasons.

In spring, cut back all stems to around 60cm (24in) from the base, aiming to make your cuts above a healthy fat bud or shoot. Thin out the vigorous new growth as it develops, and train it back on to the support.

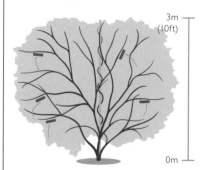

Reduce flowered shoots after flowering if not growing for fruit as well.

MOCK ORANGE *PHILADELPHUS*

This hardy, medium-sized shrub, with its dark green foliage, is a garden favourite thanks to its abundant displays of cup-shaped, white flowers in late spring and summer. These delicate, four-petalled blooms are intensely fragrant, with an intoxicating sweet scent that can fill a whole garden.

PLANT TYPE Deciduous shrubs
HEIGHT Up to 3m (10ft)
SPREAD Up to 3m (10ft)
FLOWERS ON Last year's stems, in late spring–summer
LEAF ARRANGEMENT Opposite
WHEN TO PRUNE After flowering
RENOVATION Yes

HOW THEY GROW

Mock orange is usually erect or arching in habit, and is easy to grow, making it perfect as a specimen shrub or for the back of a border. It can be planted in a row as an informal screen or hedge, and has the double bonus of being attractive to bees as well as proving rabbit-resistant. Grow in any moist but well-drained soil, except chalk, and avoid waterlogged sites. It tolerates partial shade but flowers better in sun.

Most mock oranges are fully hardy down to –15°C (5°F) and suitable for cool-temperate climates, apart from half-hardy *P. mexicanus* and frost-hardy *P. maculatus* 'Mexican Jewel', which is hardy down to –5°C (23°F) and will benefit from a sheltered sunny position in the garden.

The most popular cultivars include compact-growing *P.* 'Belle Étoile', which bears white single flowers with pale purple centres, and *P. coronarius*, with its creamy blooms. *Philadelphus* 'Virginal' has show-stopping, double flowers all along its stems, while *P. coronarius* 'Aureus' develops striking yellow foliage. *Philadelphus* 'Manteau d'Hermine' is a good choice for a small garden as it grows to just 80cm (32in) high.

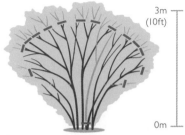

Cut back flowered shoots and thin out some of the oldest stems.

HOW TO PRUNE

Mock oranges do not always require annual pruning, but you can keep their growth compact and vigorous by giving them some attention every other year at least. After flowering, shorten flowered shoots back to strong fresh growth. Remove dead, diseased, and damaged stems, and thin out the centre of each shrub by cutting out about one in three of the oldest stems at the base.

RENOVATION PRUNING Neglected shrubs can become congested in the centre and bare at the base of stems over time, and carry fewer flowers. Fortunately, mock oranges respond to hard pruning, although they will not bloom for a year, or two, after being renovated because they flower on the previous year's growth.

After flowering, cut all stems down to 30cm (12in) from the ground. The following year, select the strongest new stems and allow these to develop, and prune out the remaining ones.

In spring, *Philadelphus* 'Belle Étoile' produces highly fragrant white flowers marked pale purple in the centres.

PHOTINIA *PHOTINIA*

Most photinias for the garden are evergreens chosen for their striking, bright red, fresh shoots, which contrast beautifully against the older, dark green leaves. When left unpruned, they also bear white spring flowers, with a slightly unpleasant odour, followed in some varieties by berries.

PLANT TYPE Deciduous and evergreen shrubs
HEIGHT Up to 4m (13ft)
SPREAD Up to 4m (13ft)
FLOWERS ON Last year's and current year's stems, in spring
LEAF ARRANGEMENT Alternate
WHEN TO PRUNE Late winter and in late spring or early summer; hedges and shaped shrubs in late winter–midsummer
RENOVATION Yes

HOW THEY GROW

This dense, fast-growing shrub develops an upright, rounded, or spreading habit. Deciduous types have good autumn colour, but it is the evergreen types such as the vigorous hybrid *P.* × *fraseri* 'Red Robin' that are mostly widely grown and used particularly as hedging as well as clipped shapes, standards, and specimen shrubs. Early in the season, the fresh new shoots and leaves provide a welcome pop of scarlet against the glossy older foliage; the leaves turn green as they mature.

Most garden types are fully hardy, though evergreens may experience leaf drop and frost damage in exposed areas, so benefit from a sheltered position such as at the base of a warm, south-facing wall. The majority grow in any fertile, moist but well-drained soil, although very heavy clay soils should be improved with organic matter to ensure success as photinia dislikes waterlogged or wet winter conditions. The best young leaf colour and flowering occur in full sun, but photinias also grow in partial shade.

HOW TO PRUNE

FREE-GROWING TYPES Most free-growing specimen photinias require minimal pruning, aside from the removal of dead or damaged growth and wayward or crossing stems in late winter. However, you can keep them to size and growing in a dense, bushy, more compact manner, and maximize the flushes of young red growth on

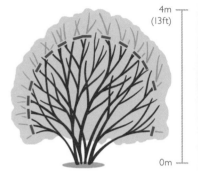

4m (13ft)

0m

Trim stems on evergreens to encourage colourful fresh growth.

evergreen bushes such as 'Red Robin', by shortening stems by about 15cm (6in) in late spring or early summer.

HEDGES AND SHAPED SHRUBS These can be trimmed up to three times a year. Try to do this tidying job before it gets too late in the season – that is, not after midsummer or into autumn – in case the new growth gets damaged by early frosts.

RENOVATION PRUNING Photinias can get leggy and bare towards the base of stems, but do respond to hard pruning. To renovate an overgrown specimen or hedge, cut back stems to about 60cm (24in) from the ground in late spring or early summer, making the cuts above outward-facing buds. Thin out the resulting fresh growth, keeping the strongest stems. Photinias shaped as topiary or standards are not suitable for renovation pruning.

Young leaves on evergreen *Photinia* × *fraseri* 'Red Robin' are red-tipped.

The berries of *Photinia davidiana* develop from its white spring flowers.

CAPE FUCHSIA *PHYGELIUS*

These wonderful but rarely grown garden plants are so called because of their similar appearance to fuchsias at first glance. They are grown for their long, large, showy panicles of tubular flowers in warm-coloured shades, which, if deadheaded continuously, are produced over a long period.

PLANT TYPE Evergreen and semi-evergreen shrubs and subshrubs
HEIGHT Up to 1.2m (4ft)
SPREAD Up to 1.2m (4ft)
FLOWERS ON Current year's stems, in summer–autumn
LEAF ARRANGEMENT Opposite; sometimes alternate
WHEN TO PRUNE Spring
RENOVATION No

Red buds open into dark pink flowers on *Phygelius × rectus* 'Devil's Tears'.

HOW THEY GROW

Bushy vigorous cape fuchsias are mostly small, evergreen to semi-evergreen shrubs and subshrubs with an erect suckering habit. Although mostly frost hardy to between −5°C (23°F) and −10°C (14°F), they are typically grown as herbaceous perennials in climates where the temperatures drop below 0°C (32°F) on a regular basis. They are

Phygelius × rectus **'Moonraker'** is a fast-growing, upright, suckering shrub.

easy to grow, are perfect for borders, and are often grown in containers. In their native South Africa, cape fuchsias grow in wet areas in the wild, and there are only two species, but due to breeding there are now many hybrids and cultivars, too.

In gardens, cape fuchsias are not particularly fussy about their growing conditions, and they thrive in any moist but well-drained soil, in full sun. In cool-temperate climates, they appreciate a sheltered position out of cold drying winds. Whether treating cape fuchsia as a shrub or a herbaceous perennial, apply a mulch overwinter to protect the roots.

Phygelius × rectus cultivars such as pale yellow-flowered 'Moonraker' or dark pink-blooming 'Devil's Tears' are some of the most popular. The species *P. capensis* bears reddish orange flowers

and is one of the toughest types, being hardy to −10°C (14°F). All cape fuchsias should be deadheaded continuously throughout summer to encourage the development of more blooms.

HOW TO PRUNE

When grown as a shrub, cape fuchsia can be given a light prune during spring, once growth begins again, to retain its shape. Trim back any overlong or wayward shoots that spoil the symmetry of the plant, and thin out old congested stems. Established plants grown in ideal conditions may produce suckers, which should be removed as soon as they are spotted.

In cool-temperate climates, where cape fuchsias are likely to experience frost damage, treat them like herbaceous perennials by cutting back all stems hard near their bases in spring.

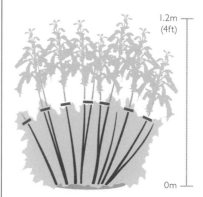

1.2m (4ft)

0m

Deadhead flowerheads throughout the summer to prolong flowering.

PIERIS *PIERIS*

These compact shrubs offer year-round interest, with their usually glossy, leathery, dark green leaves. They pack an extra punch in spring with their bright new growth as well as panicles of small flowers, like rows of tiny bells. When planted in the right conditions, pieris are low-maintenance.

PLANT TYPE Evergreen shrubs
HEIGHT Up to 4m (13ft)
SPREAD Up to 3m (10ft)
FLOWERS ON Last year's stems, in spring
LEAF ARRANGEMENT Alternate or whorled
WHEN TO PRUNE After flowering
RENOVATION Yes

Pieris formosa var. forrestii **'Wakehurst'** performs best when grown in moist but well-drained, acidic soil in dappled shade.

HOW THEY GROW

Pieris are compact, mostly rounded or upright, bushy shrubs. They are grown for their dazzling display of fresh, vibrant, red or pink leaves and shoots that appear at the ends of stems in late winter and early spring. This is followed by clusters of small, white or sometimes pink, urn-shaped flowers, similar to those of lily-of-the-valley (*Convallaria*), in mid-spring. Often used in woodland gardens, they will also bring a pop of colour to a shrub border.

Although pieris are hardy, they appreciate a sheltered position and a mulch to protect the roots in winter. They do best if planted where there is shade in the morning, to avoid frost damage to the foliage, but can be grown in full sun or partial shade, in any aspect except north-facing. The most important fact is that they must be planted in acidic soil, which is moist but well-drained. Pieris will not tolerate chalky or alkaline conditions, and soil with a high pH will need its acidity increased by the addition of composted pine needles or ericaceous compost, if this shrub is to thrive.

Popular cultivars include *P.* 'Forest Flame', which is a large type to 4m (13ft) with white flowers and fiery young growth. The much smaller *P.* 'Flaming Silver', which grows to just 1.5m (5ft), has variegated, white-edged leaves.

HOW TO PRUNE

Pieris require minimal pruning. After flowering, deadhead spent flower clusters and cut any damaged and dead stems back to healthy growth. Trim any overlong stems that are ruining the shape of the plant, and any wayward or leggy shoots that affect its natural symmetry.

RENOVATION PRUNING You can reshape pieris by cutting back hard into old wood, even down close to the ground. You can also completely renovate an overgrown specimen in mid-spring, just as growth begins, by chopping down all stems to 30–60cm (1–2ft) from the base. This will result in the loss of the flowers in that season and the following year. After renovation, mulch around the base and feed the plant, and keep it well watered throughout the summer.

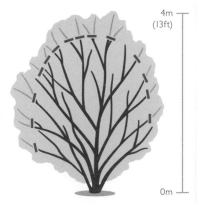

4m (13ft)

0m

Remove the sprays of faded flowers once blooming has finished.

CHERRY *PRUNUS*

Ornamental cherries are a broad group in which the deciduous trees are generally grown for their beautiful displays of spring blossom and colourful autumn leaves, while the evergreen shrubs are typically used as hedging to offer cover year-round. All are invaluable additions to the garden.

PLANT TYPE Deciduous and evergreen trees and shrubs
HEIGHT Up to 10m (33ft)
SPREAD Up to 5m (16ft)
FLOWERS ON Last year's stems, in spring, summer, or autumn–spring
LEAF ARRANGEMENT Alternate
WHEN TO PRUNE Trees, shrubs, and deciduous hedges after flowering; evergreen hedges in early–mid-spring
RENOVATION Evergreens only

HOW THEY GROW

Deciduous flowering cherry trees vary in growth habit from those with open spreading canopies like great white cherry (*P.* 'Tai-haku') to columnar forms like that of *P.* 'Amanogawa' and weeping types like *P.* 'Pendula Rosea'. They also differ widely in size, from miniature *P.* 'Kojo-no-mai', which can be grown in a pot, to bird cherry (*P. padus*), which can reach up to 15m (49ft). Popular ornamental types include *P.* 'Accolade', which has pale pink, semi-double blossom and good fiery autumn foliage colour, and *P.* 'The Bride', which carries single white flowers with red anthers. *Prunus* × *subhirtella* 'Autumnalis' produces its blossom from autumn to spring, and has bronze new leaves. Tibetan cherry (*P. serrula*) is grown for its rich reddy brown bark in winter.

Prunus **'The Bride'** can reach a height and spread of 8m (26ft).

Shrubby evergreen species including cherry laurel (*P. laurocerasus*) and Portuguese laurel (*P. lusitanica*) have a dense bushy habit and are excellent for growing as hedges or screening and as ground cover. They bear glossy, dark green leaves and long racemes of white flowers in spring or summer.

Cherries grow in any moderately fertile, moist but well-drained soil. The deciduous flowering trees do best when planted in full sun in an open position but sheltered from strong winds, while the evergreen shrubby types grow in full sun or partial shade.

HOW TO PRUNE

DECIDUOUS TREES Ornamental flowering cherries do not require regular pruning every year and are best left alone, unless intervention is essential. However, if it is necessary to remove dead, diseased, damaged, rubbing, or crossing branches, you should prune during the growing season, after flowering, in summer, to avoid problems with disease. At the same time, thin out congested growth in the centre of the canopy, if required and it is safe to do so, and cut back any growth that spoils the shape of the tree to healthy growth.

EVERGREEN SHRUBS Evergreen bushes benefit from an annual trim after flowering. Shorten flowered stems back to just above a healthy leaf node, and trim back any growth that ruins the form of each shrub.

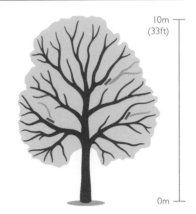

10m (33ft)

0m

Cut out damaged and dead wood on deciduous trees after flowering.

HEDGES Deciduous hedges can be trimmed after flowering, with hedge shears or trimmers. Clip evergreen hedges between early and mid-spring. Since flowering cherry and Portuguese laurel both have large leaves, it is advisable to use shears or secateurs rather than hedge trimmers, to avoid cutting through individual leaves.

RENOVATION PRUNING Deciduous flowering cherry trees will not respond well to hard pruning, so it is best to carefully choose the right tree for your size of garden; then, growth restriction will not be necessary. Shrubby evergreen types, however, can be cut back hard, even into old wood, to renovate an overgrown specimen. After flowering, in late spring or early summer, reduce stems to the required size or shorten to 30–60cm (12–24in) from the base.

FIRETHORN *PYRACANTHA*

Masses of fragrant white flowers appear from spring onwards, but firethorn is commonly grown for its autumn berries in vivid shades of red, orange, and yellow, which can last through winter. It also provides bright or dark green leaves year-round and is of great benefit to wildlife.

PLANT TYPE Evergreen small trees and shrubs
HEIGHT Up to 4m (13ft)
SPREAD Up to 4m (13ft)
FLOWERS ON Last year's stems, in late spring–midsummer
LEAF ARRANGEMENT Alternate
WHEN TO PRUNE Freestanding shrubs after flowering; wall-trained shrubs and hedges in late spring–late summer
RENOVATION Yes

HOW THEY GROW

These spiny evergreen shrubs have various growth habits: *P. coccinea* 'Red Cushion' is low and spreading, while *P.* 'Orange Glow' is upright and *P. rogersiana* 'Flava' has arching growth. All produce hawthorn-like, cream-coloured blossom from late spring to midsummer, followed by clusters of rounded berries that appear in late summer and ripen during autumn in bright, eye-catching hues.

Firethorn can be grown as a freestanding specimen tree or shrub or be trained against a fence or wall. It makes an excellent barrier or boundary hedge thanks to its dense-growing, spiked branches, which provide privacy, security, noise reduction, and a windbreak. It is wildlife friendly, too, particularly for birds, which use it for safe nesting and shelter, and also enjoy the berries as a food source. Firethorn is fully hardy and thrives in any moist but well-drained soil, in full sun or partial shade, in any aspect – even by north- or east-facing walls – but will benefit from a sheltered position with protection from cold drying winter winds.

HOW TO PRUNE

When pruning firethorn, make sure to wear protective equipment, especially thick gloves, to prevent injury from the long, very sharp thorns.

FREESTANDING SHRUBS Prune firethorn minimally so that you can enjoy the flowers and berries through the year. After flowering, remove dead, diseased, and damaged growth, as well as any crossing and wayward stems.

Prune any awkward growth and weak, damaged, or crossing stems.

WALL-TRAINED SHRUBS After flowering, trim back sideshoots to two or three clusters of developing berries. Remove inward- and outward-growing stems and any other unwanted growth. Also cut out unproductive bare stems. Tie in new shoots as needed. In spring, make way for fresh growth by removing the old spent berries.

HEDGES Between late spring and late summer, you can clip hedges several times to keep them in shape. Usually, two or three trims a year will be needed for a formal finish.

RENOVATION PRUNING Firethorn responds to hard pruning, but renovation can leave them vunerable to problems like scab and fireblight, so this should be borne in mind if cutting a plant back drastically. Cut all stems, in sections, back to the required size – right down to 30–60cm (12–24in) from the ground if necessary. As they flower on old wood, firethorn won't put on as good a floral show or fruit as well during the year after renovation.

Firethorn's bright berries bring colour to the winter garden.

These small flowers adorn *Pyracantha* 'Watereri' during late spring.

RHODODENDRON *RHODODENDRON*

Rhododendrons put on an unforgettable spring show, with innumerable blazing blooms in shades of red and pink, white, purple, orange, and yellow smothering these plants and lighting up shady woodland walks. They also offer either smart green foliage year-round, or attractive autumn colour.

PLANT TYPE Deciduous and evergreen trees and shrubs
HEIGHT Up to 3m (10ft)
SPREAD Up to 3m (10ft)
FLOWERS ON Last year's stems, in late autumn–late summer
LEAF ARRANGEMENT Opposite or whorled
WHEN TO PRUNE After flowering
RENOVATION Yes

Rhododendron luteum is an azalea bearing strongly scented, spring blooms.

Compact *Rhododendron* 'Grumpy' grows to about 1m (3ft) across.

HOW THEY GROW

There are hundreds of species and many hybrids of rhododendron, and they vary considerably in leaf and flower size and in form as well as in growth habit, from low ground-cover plants and rounded shrubs to huge, tree-like forms that reach up to 25m (80ft) high. However, most garden cultivars grow to an eventual height and spread of around 3m (10ft). Try to choose a plant that is a suitable size for the space you have, as pruning to restrict the growth of rhododendrons is not generally advisable.

The range of rhododendron types includes evergreen and deciduous ones, as well as those known as azaleas. Evergreen rhododendrons such as *R. barbatum* have bright showy blooms in a multitude of colours, and glossy green leaves year-round, while the deciduous types like azalea *R. luteum* bear funnel-shaped or tubular flowers

that are occasionally scented, and sometimes also produce eye-catching autumn leaf colour. The majority of rhododendrons flower in spring, but some bloom earlier, in winter, while others put on a show as late as autumn.

Rhododendrons are available as shrubby hybrids (for example, pink *R.* 'Roseum Elegans'), which are perfect for woodland gardens, and as compact mounding plants that are small enough to grow in a mixed border or pot (for example, *R. impeditum*, which has aromatic foliage) and as dwarf types like *R.* 'Princess Anne'.

Evergreen rhododendrons are mostly hardy but can suffer from frost damage, so appreciate a sheltered position out of cold drying winds and away from frost pockets and early morning sun. Deciduous types are fully hardy. Both prefer growing in partial shade in acid soil that is moist but well-drained. In general, they like temperate conditions with high rainfall.

HOW TO PRUNE

Rhododendrons don't require much pruning beyond the occasional removal of dead, diseased, and damaged wood, and any wayward growth that ruins the shape of the plant. Do this after flowering, if necessary, when you can also deadhead the spent flowers.

RENOVATION PRUNING Rhododendrons can be cut back hard to shape or rejuvenate them, but may not always respond well. There is a greater chance of success with deciduous types, especially azaleas, which can be taken back up to 30cm (12in) from the base if required, in winter.

Renovate overgrown evergreen rhododendrons in stages, cutting back one of the main stems to 30cm (12in) from the base, after flowering. See how the plant responds before proceeding to shorten other stems over a few seasons. Plants may not flower for several seasons after drastic pruning.

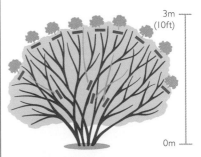

3m (10ft)

0m

After flowering, remove spent flowerheads and any awkward stems.

SUMACH *RHUS*

Although it bears interesting flowers and fruit, sumach is grown predominantly for its impressively long, fern-like foliage, made up of many leaflets, which turn brilliant shades of yellow, red, and orange in autumn. It is very easy to grow and care for, but can spread vigorously.

PLANT TYPE Deciduous trees and shrubs
HEIGHT Up to 5m (16ft)
SPREAD Up to 6m (20ft)
FLOWERS ON Last year's and current year's stems, in spring–summer
LEAF ARRANGEMENT Alternate
WHEN TO PRUNE Late winter–early spring
RENOVATION Yes

The dark red, conical fruits show particularly well in winter.

HOW THEY GROW

Stag's horn sumach (*R. typhina*) has an upright form and a spreading habit, and is often grown as a small, occasionally multi-stem, tree. As well as its striking long leaves and fiery autumn foliage colours, it also produces upright, cone-shaped panicles of greenish yellow flowers in spring and summer, which on female plants develop into clusters

Impressive fiery shades develop in stag's horn sumach's autumn leaves.

of downy red fruit. The browny red, velvety shoots have a soft feel, and branch densely in a way that is reminiscent of deer antlers, which is how this plant gets its unusual name. Cut-leaved stag's horn sumach (*R.t.* 'Dissecta'), at just 2m (6½ft) tall, has a shrubbier form and finely cut foliage.

Sumachs are grown in woodland gardens, shrub borders, and as specimen plants, and are particularly good for sloping sites. They are fully hardy and like an open position in full sun, in any moist but well-drained soil that is moderately fertile. They produce suckers, which spread to form colonies, and, depending on growing conditions, they can become invasive.

HOW TO PRUNE

Always wear gloves when pruning, as this plant's sap can aggravate skin allergies. For freestanding specimens with the space to grow to their natural size and shape, only basic pruning is necessary. In early spring, remove dead, diseased, and damaged growth, and any crossing or rubbing stems. Also pull away unwanted suckers at the base, to control their spread. To ease congestion in a mature plant, you can cut one-third of the oldest branches back to the ground.

If space is an issue and growth must be controlled, you can prune sumach hard each year, cutting it back within 30–60cm (12–24in) of the ground during late winter.

RENOVATION PRUNING To renovate overgrown plants or colonies, in late winter, shorten all stems back to 30–60cm (12–24in) long.

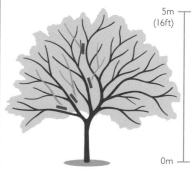

Remove crossing and dead growth in early spring on freestanding trees.

FLOWERING CURRANT *RIBES*

Although the genus *Ribes* is well known for plants with edible fruits, such as currants and gooseberries, it also contains many ornamental shrubs grown primarily for their showy spring flowers in shades of white, yellow, and pink. Some deciduous types produce well-coloured autumn foliage, too.

PLANT TYPE Deciduous and evergreen shrubs
HEIGHT Up to 2.5m (8ft)
SPREAD Up to 2.5m (8ft)
FLOWERS ON Last year's stems, in spring–summer
LEAF ARRANGEMENT Alternate
WHEN TO PRUNE After flowering
RENOVATION Yes

HOW THEY GROW

Most flowering currants are upright and spreading shrubs with a bushy habit, and, unlike those grown for fruit, are usually spineless. They vary in size from *R. laurifolium*, a dwarf evergreen that reaches 1m (3ft) and produces yellow flowers, to *R. sanguineum* and its cultivars, which grow to 2.5m (8ft) and have scented pink flowers. The spring blooms are small and bell-like, and tend to hang down in little clusters or racemes, and are followed by berry-like fruit in various colours. Deciduous *R. americanum* develops the best autumn foliage colouring.

Flowering currants can be used in the garden everywhere from the rock garden to a shrub border or as a specimen shrub, and look particularly good when planted alongside spring-flowering forsythia. They grow in any moist but well-drained soil in full sun, in any aspect, and are mostly fully hardy, though *R. speciosum*, an upright deciduous type with red flowers, is slightly more tender and benefits from protection in areas that experience hard frosts; it can be trained against a warm wall. *Ribes alpinum*, on the other hand, is very hardy and tough, and puts up with some shade and poor soil.

HOW TO PRUNE

These shrubs are quite vigorous and benefit from an annual prune.

FREESTANDING SHRUBS After flowering, remove any dead, diseased, or damaged wood, and shorten flowered stems back to a healthy bud lower down. Encourage fresh growth from the base by removing up to one-quarter of the stems, starting with the oldest. Some gardeners also cut the remaining stems back hard to whatever size they wish, to keep the shrubs to their allotted space.

WALL-TRAINED SHRUBS Cut back all flowered shoots to within two to four buds of the permanent framework, after flowering, and remove all inward- and outward-growing shoots.

RENOVATION PRUNING Some types of flowering currant, such as *R. sanguineum*, are prone to becoming leggy if not regularly pruned. Large overgrown bushes can be renovated after flowering by cutting all stems down to 30cm (12in) from the base.

Ribes sanguineum **'Pulborough Scarlet'** carries spring blossom.

The fuchsia-like blooms of *Ribes speciosum* are borne on prickly stems.

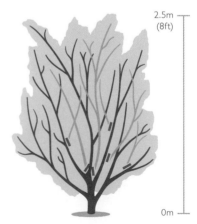

2.5m (8ft)

0m

Reduce flowered shoots and cut out some old stems after flowering.

ROSE *ROSA*

The rose has captured our hearts for centuries with its memorably scented summer blooms. These iconic plants come in a broad variety of forms, from rambuctious ramblers to border bushes. The thorny stems are as famous as the flowers, which are often followed by luscious hips.

PLANT TYPE Deciduous and semi-evergreen shrubs and climbers
HEIGHT Up to 10m (33ft)
SPREAD Up to 10m (33ft)
FLOWERS ON Last year's and current year's stems, in summer
LEAF ARRANGEMENT Alternate
WHEN TO PRUNE After flowering, or late winter
RENOVATION Yes for most types

ROSE CLASSIFICATIONS

There are thousands of roses, and it can be daunting trying to figure out the different types. In general, roses can be classified either as Species or as cultivars (which are subdivided into Old roses and Modern roses).

SPECIES ROSES We might think of these as wild roses. They are large plants with simple single flowers, and they tend to bloom once, in early summer, and then develop hips. *Rosa spinosissima*, with its single, creamy white blooms, is a typical Species rose.

OLD ROSES These were typically bred and introduced up to the mid-nineteenth century. This huge group includes Bourbons, China roses, Centifolias, Damasks, Noisettes, Albas, Gallicas, Rugosas, and Tea roses. Typical Old roses are white-bloomed Damask rose R. 'Madame Hardy' and pink Centifolia R. 'Fantin-Latour'.

MODERN ROSES These include more recently bred varieties such as Hybrid Teas, English roses, Floribundas, Miniatures, Polyanthas, Patio roses, and Ground-cover or Carpet types. Many are remontant (repeat-flowering), like R. Gertrude Jekyll, an English rose with highly scented, deep pink, double flowers in summer.

A rose's classification is one of the determining factors in how it should be pruned. Other factors include how frequently it flowers and its form (whether it is a shrub or bush rose, a climber or rambler, a ground-cover or miniature rose).

Rosa **'Geranium'** (*moyesii* hybrid) has big long hips after its deep red flowers.

Hybrid Tea rose POETRY IN MOTION has fragrant blooms in summer and autumn.

MOTION and pink R. 'Ballerina' are good for mixed borders or a rose garden. Climbers such as R. 'Madame Alfred Carrière', which often repeat-flower, are trained up and over structures and to cover surfaces. The more vigorous, typically once-flowering ramblers, such as R. 'Rambling Rector', are often grown up into trees to give a wilder romantic feel. Some roses like R. *spinosissima* are used as hedges, and others like R. Flower Carpet Series are great for ground cover. Roses such as R. ICEBERG or R. BONICA can be trained as standards, while smaller types such as red-bloomed Patio rose R. PETER PAN work well in containers.

Flowers also differ greatly in form and colour, ranging from simple, five-petalled, flat or cupped blooms to complex, rosette-shaped, double flowers, in every shade from red and

HOW THEY GROW

Species roses are usually climbers or shrubs, while Old and Modern roses include climbing and rambling forms. Other roses within each classification vary widely in growth habit, from upright to arching and trailing or low spreading types. Some are vigorous or suckering, and others are compact or lax. Ramblers like R. 'Paul's Himalayan Musk' can reach up to 10m (33ft). Miniatures like R. 'Lavender Jewel' can be as small as 25cm (10in). Roses can be used in many different ways. Cultivars like yellow R. POETRY IN

TOP TIP ROSES ARE USUALLY AVAILABLE AS EITHER POT-GROWN OR BARE-ROOT SPECIMENS. YOU CAN PLANT POT-GROWN ROSES ANY TIME OF THE YEAR, WHEREAS BARE-ROOT ROSES WILL BE DISPATCHED ONLY IN WINTER AND SHOULD BE PLANTED AS SOON AS POSSIBLE – THOUGH NOT IF THE GROUND IS FROZEN.

white to apricot, yellow, pink, and purple. Rose fragrance can be fruity, musky, tea, myrrh, floral, or the distinctive, Old rose one. The hips that follow the flowers are another great attribute, with some of the best examples on *R.* 'Geranium' and *R. villosa* subsp. *villosa*.

Although the sheer array of types, forms, and classifications can be confusing, caring for roses is straightforward, because they are hardy and easy to grow, and do not need much maintenance. Plant in an open position in full sun, in any moderately fertile, moist but well-drained soil as long as it is not waterlogged or too dry. On heavy clay soils, improve drainage with grit or gravel, and organic matter. On poor sandy soils, dig in compost and well-rotted manure before planting.

HOW TO TRAIN

DOMED SHRUBS To create an attractive domed or fountain shape to your rose, which will also encourage more flowers all along the stems, you can train it over a ready-made metal dome support, or over a dome made of hazel rods (stems) bent into semicircles and inserted in the ground. After the plant has come into growth in spring, when the stems are long enough, arrange them one by one over the support or rods, and tie in place. Alternatively, you can pull each stem down to the ground in an arc and pin down in the soil.

WEEPING STANDARDS To create a weeping standard, start with a standard rose attached to a strong stake to keep its stem straight. After planting, allow the plant to establish for two years. Once its shoots are long enough to manipulate, attach a length of garden twine to each shoot, and carefully pull each one downwards, tying the end of the string to the stake. Long-flowering *R.* 'New Dawn' makes an excellent large weeping standard.

CLIMBERS AND RAMBLERS These roses need to be tied into supports and structures because they are not self-clinging. They flower more and grow better when stems are forced down to near horizontal, so, whatever

Old rose 'Louise Odier' looks good when trained over a dome made of hazel.

the support, the training idea is to bend, curve, arch, and force the stems down, rather than letting them grow naturally straight up. Some gardeners prefer to train stems into a series of loops before tying in. This last technique is best restricted to roses that have very lax stems, and is easiest and safest with flexible new growth, which can be encouraged with regular pruning. Trellis or a system of strong wires will support a rose grown against a wall or fence. Let the plant establish and put on growth for a year or so after planting, and then begin to train its growth by pulling the stems down and tying them in along the trellis or wires rather than up it. On obelisks, wigwams, and pillars, spiral the stems around the support, gently forcing the stems horizontally and tying in as they grow.

Scented double flowers are carried on rambling rose 'Alexandre Girault'.

4m (13ft)

0m

To train a rambling rose, tie stems horizontally along their support; also reduce sideshoots after flowering.

HOW TO PRUNE

Roses require various pruning methods at different times of year, but there are some guidelines that apply to a majority of types. Standard roses should be pruned according to the type of rose that forms the head of the standard. When pruning, always wear gauntlets to protect your hands and arms from the sharp thorns.

Immediately after planting, all roses except shrub, climbing, and standard ones can be pruned hard back to 15–30cm (6–12in) from the ground, to encourage fresh vigorous growth.

For established plants, prune every year and always begin by cutting back dead, diseased, damaged, and weak stems, and crossing or rubbing growth. Those being pruned in late winter should have any remaining foliage removed and then burned or binned; never compost such leaves, to avoid spreading disease.

The aim for most rose types is to develop a plant with an open centre, with plenty of fresh vigorous growth from the base to replace older stems.

Clusters of myrrh-scented, pale golden to apricot flowers are borne throughout summer on compact bush rose ABSOLUTELY FABULOUS.

Some gardeners like to keep stems for a maximum of three years before they are deemed too old and unproductive, and so need cutting out at the base.

Many roses are grafted, so look out for suckers growing from beneath the graft union. Find where they connect to the base of the plant and remove them by pulling off (see p.23).

SHRUB ROSES For shrub roses including Species roses, Old roses that flower once (for example, Alba, Centifolia, Damask, Gallica, Moss, Scots, and Sweet Briar types) and Modern roses that flower once (for example, Hybrid Musk and Rugosa types), prune lightly after flowering during late summer. As well as doing the basic removal of dead and crossing growth described above, if necessary you should ease congestion in the centre of the shrub to keep it open, by taking out one or two of the older stems. It is advisable not to restrict the natural size and form of Modern once-flowering shrub roses, but, if some neatening is required, shorten sideshoots by one-third. Never deadhead when you want hips to develop.

For repeat-flowering shrub roses including Bourbon, China, and Portland types, prune in late winter. Cut back the previous season's growth by up to one-third, and reduce sideshoots back to two or three buds. You can also remove one or two of the older stems completely. To encourage more flowers, deadhead throughout the summer months.

RENOVATION PRUNING Once-flowering shrub roses can be renewed by cutting out up to one-quarter of the flowered stems back to the base. Repeat-flowering shrub roses that have got leggy or too large can be cut back by one-half and have one-quarter of the oldest stems taken out.

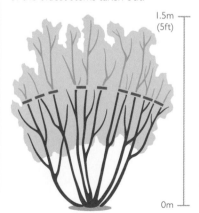

1.5m (5ft)

0m

On repeat-flowering shrub roses, reduce growth by up to one-third.

Always prune out all dead and damaged stems on every type of rose.

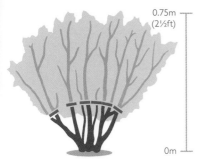

0.75m
(2½ft)

0m

Cut the stems of Floribunda bush roses to 30cm (12in) from the ground.

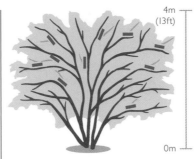

4m
(13ft)

0m

On climbers, shorten sideshoots to three or four buds in winter.

BUSH ROSES Prune bush roses, such as Hybrid Tea and Floribunda types, in late winter. Hybrid Teas can be cut quite hard, removing old and weak growth and shortening the most vigorous stems to 15cm (6in) from the base. Reduce the remaining shoots to 10cm (4in) from the base.

Floribundas are pruned slightly less. Cut back their most vigorous stems to 30cm (12in) from the base and the remaining stems to within 15cm (6in). Every few years, shorten some older stems to within 10cm (4in) of the base.

Patio and Polyantha roses can be pruned in the same way as Floribundas, but a little less severely. Reduce the main stems by about one-third, and remove any overlong or wayward shoots that spoil the shape entirely.

RENOVATION PRUNING To rejuvenate bush roses, cut out up to one-third of the oldest stems every year for three years.

CLIMBERS After planting a climber, and for the first two years in winter, remove only damaged, diseased, or dead growth, and tie in new shoots to the support. Once established, keep growth to within the allotted space and reduce sideshoots back to three or four buds. If necessary, cut out old unproductive stems close to the base. Deadhead in the growing season to encourage repeat-flowering.

> **TOP TIP** IN GENERAL, A CLIMBING ROSE WILL REPEAT-FLOWER THROUGH SUMMER, BUT A RAMBLER WILL BLOOM ONLY ONCE, IN EARLY SUMMER.

RENOVATION PRUNING To reinvigorate a climber, cut the oldest woody stems back to the ground. This can be done in stages, removing one or two stems a year over three years, if preferred. Retain about six of the newer vigorous stems but shorten their sideshoots.

RAMBLERS After planting, prune a rambler by cutting all its stems back to 40cm (16in) from the base, to encourage vigorous new growth. For the first two years, train stems horizontally on to the support (see p.38), and shorten sideshoots back to two or three buds after flowering. Once established, prune after flowering in late summer,

Trim back shoots of once-flowering shrub roses as they fade, to tidy them

reducing the sideshoots to three or four buds and cutting back any stems that grow beyond the desired space. Every few years, remove one or two of the oldest stems close to the base.

RENOVATION PRUNING Renovate an overgrown rambler in late autumn or winter, cutting out the oldest woody stems completely. Keep up to six of the most vigorous new ones, tying them in as replacements.

GROUND-COVER ROSES Prune Ground-cover or Carpet roses in late winter. They do not need much regular attention beyond basic pruning (see *opposite*), but if they outgrow their allotted space cut back any overlong stems and wayward growth that spoils the symmetry of the plant. Shorten the most vigorous shoots by up to one-third, and cut sideshoots back to two or three buds.

RENOVATION PRUNING If a ground-cover rose becomes overgrown and twiggy, it can be cut back drastically to 10cm (4in) from the base in late winter Alternatively, to rejuvenate, remove up to one-quarter of the oldest stems.

MINIATURE ROSES Dwarf roses don't require much pruning other than basic removal of dead, diseased, damaged, and weak growth in late winter.

Rosa **FLOWER CARPET WHITE** makes excellent mounding ground cover.

ORNAMENTAL BRAMBLE

RUBUS

Ornamental brambles are in the same genus as raspberries and blackberries. They are cultivated mostly for their saucer-shaped flowers and colourful, sometimes edible fruit, or the decorative prickly stems on some deciduous types. Evergreen types make great ground cover.

PLANT TYPE Deciduous and evergreen shrubs
HEIGHT Up to 6m (20ft)
SPREAD 2–3m (6½–10ft) or more
FLOWERS ON Last year's stems, in early spring, summer, or summer–autumn
LEAF ARRANGEMENT Alternate
WHEN TO PRUNE Floral types after flowering; ornamental-stem types in early spring
RENOVATION Yes

HOW THEY GROW

Brambles vary in growth habit from low spreading evergreens to thicket-forming deciduous shrubs with arching bristly stems. They are a great choice for wild gardens and the edge of woodland ones. Many are vigorous with invasive tendencies, with an indefinite spread if left to grow unchecked. *Rubus* 'Betty Ashburner', a prostrate evergreen, is a typical example, making excellent ground cover. It reaches a maximum height of around 30cm (12in), has red bristles on its stems, and produces white flowers in summer.

Flowering raspberry (*R. odoratus*) is a suckering, spreading, deciduous type grown for its flowers, which are scented and pinky purple, borne over

Scented flowers appear in spring on thornless-stemmed *Rubus* 'Benenden'.

a long season from summer to autumn, and followed by edible but not tasty fruit. *Rubus* 'Benenden' is also popular for its single white blooms, but the showiest flowers are on *R. spectabilis* 'Olympic Double', which has hot-pink, double blooms in early spring.

Other deciduous species, such as *R. cockburnianus*, *R. thibetanus*, and *R. biflorus*, are grown mostly for the effect of their prickly sculptural stems in winter. The shapes created by the bristle-laden stems are made all the more striking by their brilliant white or grey sheen at this time of year.

The majority of these popular ornamental garden plants are fully hardy, and grow in any reasonably fertile, well-drained soil. Evergreen types and deciduous ones cultivated for their flowers can be planted in full sun or partial shade, while deciduous types grown for their winter stems need a position in full sun.

In winter, *Rubus cockburnianus* has red-purple stems with a white bloom.

HOW TO PRUNE

Let plants establish for a year or two before pruning annually. Wear gloves and protective equipment to guard against injury from the prickly stems.

FOR FLOWERS With flowering brambles such as *R. odoratus* and *R. spectabilis*, cut out one-third of the oldest stems each year after flowering. Shorten the remaining flowered stems to strong buds lower down.

FOR ORNAMENTAL STEMS Cut all stems back to the ground in early spring.

RENOVATION PRUNING To renovate any bramble, prune all stems at ground level in early spring. Those grown for flowers and fruit will not produce either for a year or two afterwards.

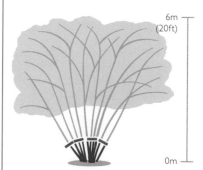

6m (20ft)

0m

Prune all stems hard on brambles with decorative stems, in spring.

WILLOW *SALIX*

This is a big diverse group of plants that are grown variously for their narrow, grey-green leaves, soft spring catkins, weeping shape, or their coloured or decoratively marked stems. Some can be woven into living willow structures, while others are harvested for basket making.

PLANT TYPE Deciduous trees and shrubs
HEIGHT 10m (33ft) or more
SPREAD Up to 8m (26ft)
FLOWERS ON Last year's stems, in spring
LEAF ARRANGEMENT Alternate
WHEN TO PRUNE Trees in late winter; shrubs in spring
RENOVATION Yes (shrubs only)

HOW THEY GROW

Willows vary from big trees to dwarf types, with weeping, upright, spreading, or prostrate habits. Those grown for their flowers, often called pussy willows, produce furry catkins on bare stems early in the year. They have both male and female catkins, but the male ones are the most prized, and are very attractive to early bees. Some of the most showy are the pinky red, silky catkins on *S. gracilistyla* 'Mount Aso', and those of *S.g.* 'Melanostachys', which open pink before changing to black.

Willow leaves are generally long and narrow, emerging grey-green in spring and turning yellow in autumn. Some of the best are on silver willow (*S. alba* var. *sericea*) and small shrubby S. 'Mark Postill'.

The weeping willow (*S. babylonica*) is known for its long trailing shoots. *Salix babylonica* 'Crispa', with spiralled curling leaves, is a good cultivar for a large garden, while *S. caprea* 'Kilmarnock' would be better for a smaller one.

Like dogwoods (*Cornus*), some willows such as *Salix alba* var. *vitellina* 'Britzensis' are coppiced for bright new stems that add colour to the winter garden (see p.28). Corkscrew willow (*S. babylonica* var. *pekinesis* 'Tortuosa') develops twisted stems that offer a sculptural effect. For a small garden, try flamingo tree (*S. integra* 'Hakuronishiki'), a dwarf type that can be grown as a standard, with variegated foliage in green and white with pink tips.

Willows are tough, hardy, low-maintenance plants that grow in any fertile, moist but well-drained soil in an open position. They prefer full sun but many tolerate some shade, and thrive in wet boggy places.

Given the right conditions, many willows will grow quite large, although they can be pollarded (see p.29). Their vigour means you should avoid planting them near buildings, walls, drains, and septic tanks, because willow roots will seek out water and can cause damage.

HOW TO PRUNE

TREES For trees with a weeping form that are small, like *S. caprea* 'Kilmarnock', prune in winter when

Flame-coloured stems adorn *Salix alba* var. *vitellina* 'Britzensis' in winter.

10m (33ft)

0m

To feature stems or foliage, coppice stems to a low framework in spring.

dormant, taking out any dead, diseased, damaged, crossing, or rubbing growth. Thin out older stems if necessary to ease congestion.

For larger trees, it is always best to employ a qualified tree surgeon for pruning or pollarding.

SHRUBS Those grown for catkins do not require regular pruning, but, if reduction is necessary, occasionally take out one-third of the stems in spring.

Willows grown for their colourful winter stems can be coppiced or pruned hard in spring, cutting back all stems to within one or two buds of the base to create a permanent low framework to cut back to every year or two. Those grown for foliage also benefit from this treatment.

RENOVATION PRUNING Shrubby willows respond well to hard pruning and can be cut back almost to the ground to help regeneration. The fresh growth often displays colourful stems.

SAGE *SALVIA*

Although many plants in the genus *Salvia* are annuals and herbaceous perennials grown for their colourful summer flowers, there are also shrubby sages, which need pruning and training. Many of these are *Salvia officinalis* cultivars with aromatic leaves used as a herb for culinary purposes.

PLANT TYPE Evergreen subshrubs, perennials, and annuals
HEIGHT Up to 1.2m (4ft)
SPREAD Up to 1.2m (4ft)
FLOWERS ON Last year's and current year's stems, in summer–autumn
LEAF ARRANGEMENT Opposite
WHEN TO PRUNE Spring
RENOVATION No

The purple young leaves on *Salvia officinalis* 'Purpurascens' age to grey-green.

Salvia 'Hot Lips' produces bicoloured blooms continuously over a long season.

HOW THEY GROW

Common sage (*S. officinalis*) is a hardy, bushy, evergreen subshrub with strongly scented, grey-green, oval leaves. It is quite drought-tolerant and is at home in Mediterranean planting schemes and herb gardens, growing to about 80cm (32in) tall. Although it does produce summer flowers atop its stems when left unpruned, common sage is generally grown for its leaves, which are used extensively, both fresh and dried, to flavour a variety of dishes. It has several attractive cultivars: *S.o.* 'Icterina' with foliage variegated in yellow and green; *S.o.* 'Purpurascens' with purple young foliage; and slightly less hardy *S.o.* 'Tricolor' with cream-edged green leaves tinted pink. There are also shrubby salvias grown for their plentiful, season-long flowers. Many hail from

Mexico and other warm places, so are tender and in cool-temperate climates should be grown in a greenhouse or conservatory. However, a few such as *S. microphylla* are frost hardy to fully hardy, depending on how exposed their position is in the garden.

Salvia microphylla 'Pink Blush' has a spreading habit and two-lipped, pink and red-purple blooms in late summer, while *S.m.* 'Rodbaston Red' is bushy with spikes of red-pink flowers from summer to autumn. Bicoloured, white and red flowers are borne on *Salvia* 'Hot Lips' from midsummer to the first frost. These shrubby ornamental salvias will act like evergreens in mild areas but may lose their stems to frost elsewhere, with new growth appearing from ground level in spring. After a very cold or wet winter, it may take until early summer for regrowth to appear.

For best results, plant sage in a south- or west-facing, sheltered spot out of cold and drying winds, in full sun. They grow in any fertile, moist but well-drained soil, except clay, because they detest wet conditions.

HOW TO PRUNE

Sage needs minimal pruning to grow well. In spring, remove any dead stems back to fresh growth and trim any shoots that are ruining the shape of the plant. How plants are subsequently trimmed depends on why they are grown. If cultivated for their flowers, deadhead shrubby sages regularly throughout the growing season to encourage more flowers.

If cultivated as a herb, harvesting the leaves regularly will function as cutting back the plant. However, if not constantly trimmed for the kitchen, prune common sage in spring to promote new leafy growth and a compact shape. Otherwise, plants lose their bushiness and develop woody leggy stems. Sage will not regrow from old wood, so replace it.

1.2m (4ft)

0m

Trim common sage regularly to keep it compact and bushy.

ROSEMARY *SALVIA ROSMARINUS*

Rosemary is a wonderful plant with many attributes, being a fragrant and flavoursome herb, an evergreen shrub, a pretty blooming bush or low hedge, and even excellent ground cover. It is drought-tolerant, unfussy about its growing conditions, and generally a great all-rounder for any garden.

PLANT TYPE Evergreen shrubs
HEIGHT Up to 2m (6½ft)
SPREAD Up to 2m (6½ft)
FLOWERS ON Last year's stems, in spring
LEAF ARRANGEMENT Opposite
WHEN TO PRUNE Shrubs in spring; hedges after flowering
RENOVATION No

HOW THEY GROW

Rosemary was previously known as *Rosmarinus officinalis* but, after discovering that the plant is actually a type of sage (*Salvia*), the name was changed to *Salvia rosmarinus*. It is valued predominantly for its densely growing, narrow, linear leaves, which are dark green on top and pale underneath. This foliage is intensely aromatic and is used most commonly for flavouring food, as a herb. However, rosemary is also an invaluable ornamental plant with pretty aromatic flowers borne in spring in shades of blue, purple, pink, and white.

Its cultivars have various growth habits from prostrate and mat-forming to upright and bushy. As well as featuring in herb and kitchen gardens, rosemary also makes a handsome evergreen specimen in a bed or border, and is also useful as low hedging.

Salvia rosmarinus 'Miss Jessopp's Upright' is an erect type with blue flowers and is one of the larger cultivars with an eventual 2m (6½ft) height and spread. *Salvia rosmarinus* Prostrata Group, on the other hand, is a low-growing variety, about 20cm (8in) tall, that makes good ground cover.

This Mediterranean shrub is quite drought-tolerant once established, and grows and flowers best in full sun. It is hardy to between −5 and −10°C (14–23°F) in most places, but can suffer frost damage in exposed gardens or during harsh winters, so benefits from a sheltered position. It prefers neutral to alkaline soil that is moist but very well-drained.

HOW TO PRUNE

If you are using your rosemary as a herb, the regular continual harvesting of the stems will promote healthy fresh growth. Like lavender (*Lavandula*), sage, and most woody herbs and subshrubs,

Rosemary has aromatic, needle-like leaves used as a food flavouring.

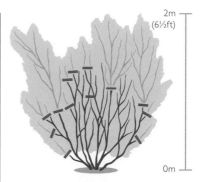

2m (6½ft)

0m

Shorten shoots to within one-third of the previous season's growth.

rosemary also benefits from regular trimming to encourage new leafy growth and a compact habit. Try to always cut back only within the last season's growth. When left unpruned, rosemary stems become leggy, unproductive, and bare at the base as well as in the centre. This plant does not reshoot reliably from old wood, so it is better to replace an old overgrown one rather than attempt renovation. Regular pruning is not required for *S. rosmarinus* Prostrata Group.

SHRUBS If growing as a specimen shrub, prune annually in spring, trimming off between 5cm (2in) and one-third of last season's growth. Also, prune one-third of the oldest stems back to a few sets of leaves.

HEDGES Prune rosemary hedges after flowering, cutting to the required shape and form for your garden.

ELDER *SAMBUCUS*

Elder is cherished for its fragrant umbels of flowers during early summer – which are often used to flavour beverages such as cordials – and for its berries. Its attractive leaves come in various colours and are finely divided. Some trees develop interesting textured bark.

PLANT TYPE Deciduous trees and shrubs
HEIGHT Up to 6m (20ft)
SPREAD Up to 4m (13ft)
FLOWERS ON Last year's stems, in early summer
LEAF ARRANGEMENT Opposite
WHEN TO PRUNE Early spring
RENOVATION Yes

HOW THEY GROW

Common elder (*S. nigra*), also known as elderberry and black elder, has an upright habit and is usually grown as a large bushy shrub or occasionally a small tree. The leaves are made up of several pairs of leaflets and can be green, almost black, yellow, variegated, and beautifully cut or dissected. Many

The white umbellifer flowers of common elder are scented and edible.

people think the foliage smells slightly unpleasant when touched. Large flat sprays of scented cream, white, or pink flowers appear in early summer, and are often picked as an ingredient in elderflower cordial or champagne. They are followed by small, sour, black, red, or white berries from late summer to autumn, which are only edible once cooked, and can be used in making elderberry wine and preserves.

The species *S. nigra* can be found growing wild in many places – its seed having escaped from gardens – and it has a reputation for becoming invasive. Better-behaved garden varieties include cut-leaved elder (*S.n.* f. *laciniata*), which has finely divided, fern-like leaves, white flowers, and purple-black berries, and golden elder (*S.n.* 'Aurea') with yellow leaves. Dark-leaved types include *S.n.* f. *porphyrophylla* 'Eva' (syn. 'Black Lace'), with its lacy, deeply cut, dark purple foliage and pinkish blooms, and *S.n.* f. *porphyrophylla* 'Gerda' (syn. 'Black Beauty'), which has burgundy-black leaves and dark pink flowers.

Elder prefers a sheltered position in any moist but well-drained soil, but it is not fussy, and some types even tolerate waterlogging. It is very hardy, down to below −15°C (5°F), and can be placed in full sun or partial shade.

HOW TO PRUNE

Allow to grow freely for two years or so after planting, removing only dead and diseased or damaged growth when

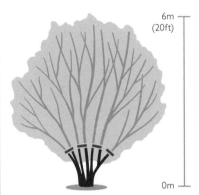

6m (20ft)

0m

Prune dark-leaved cultivars hard each year for the best foliage.

seen. From the third year onwards, you can begin to cut out a few of the oldest stems each year in early spring, to keep growth vigorous and productive of flowers and fruit, and to promote a bushy compact shape.

Elder tolerates being cut back hard to keep it to the required size, and it can be coppiced every three or four years to promote lush new foliage. With the dark-leaved cultivars, many gardeners cut the whole plant back almost to ground level every year to get the best foliage colour and effect. However, with this approach you will have to forego flowers and fruit on your plant.

RENOVATION PRUNING Elder is easy to renovate as it can be pruned really drastically right back into old wood or even down to the ground, in early spring, and it will respond by quickly producing strong, new, bushy growth.

SANTOLINA *SANTOLINA*

This Mediterranean native is a dwarf evergreen with silvery blue to grey-green leaves, which release a pleasant fragrance when touched. Its foliage colour and domed shape make santolina useful as a specimen, edging, or ground-cover plant. It has button-like, usually bright yellow blooms.

PLANT TYPE Evergreen shrubs
HEIGHT 50cm (20in)
SPREAD Up to 1m (3ft)
FLOWERS ON Last year's and current year's stems, in summer
LEAF ARRANGEMENT Alternate
WHEN TO PRUNE After flowering and in spring
RENOVATION No

Santolina chamaecyparissus **'Pretty Carroll'** produces pom-pom blooms.

HOW THEY GROW

Cotton lavender (*S. chamaecyparissus*) is a small evergreen shrub with a rounded form and compact growth habit. Its most popular attribute is its aromatic, narrow, grey-green to blue foliage, which is finely divided or toothed. It forms attractive mounds that can bring structure to a border or provide a highlight in a gravel garden, or else be shaped into a low hedge. The young shoots emerge woolly or hairy before becoming wiry stems topped with dense, bright sulphur-yellow flowers from mid- to late summer. However, many gardeners do not like these blooms and pinch them out, growing the plant only for its foliage. *Santolina chamaecyparissus* 'Lemon Queen' has paler yellow to cream flowers that are more appreciated.

There are also several interesting floral types of rosemary-leaved santolina (*S. rosmarinifolia*) including *S.r.* subsp. *rosmarinifolia* 'Primrose Gem', *S.r.* 'Lemon Fizz', and *S. pinnata* subsp. *neapolitana* 'Edward Bowles'.

Santolina is quite drought-tolerant, once established, and can cope with poor or stony soils, but grows best in moderately fertile, well-drained soil. It should be positioned in full sun in a sheltered site, because it is only frost hardy to around –5°C (23°F) and needs protection in cold winters.

HOW TO PRUNE

Prune santolina regularly to encourage fresh growth and prevent old stems from becoming leggy and bare. The centres of large old plants are also prone to splitting or dying, leaving a large hole in the centre. A trim each year will prevent this from happening.

In early spring, cut back to 2.5cm (1in) above where last season's growth meets the still-older growth. To help retain a neat shape while allowing the

Cut back to just above where the previous season's growth begins.

plant to flower, you can also deadhead the spent flowerheads or trim back flowered stems after flowering.

Santolina does not reliably regenerate from old wood, so pruning old leggy plants back hard is not an option. Most gardeners propagate or buy new plants if needed.

50cm (20in)

0m

Shorten flowered stems after flowering to keep the plant neat.

HYDRANGEA VINE *SCHIZOPHRAGMA*

Hydrangea vines are so called because they look similar to climbing hydrangea (*see p.90*), with their lacecap-style flowers in summer and dark green, heart-shaped leaves. Although slow to establish, they subsequently reward with vigorous growth to cover large structures.

PLANT TYPE Deciduous shrubs
HEIGHT Up to 8m (26ft)
SPREAD Up to 4m (13ft)
FLOWERS ON Last year's and current year's stems, in spring–summer
LEAF ARRANGEMENT Opposite
WHEN TO PRUNE Spring
RENOVATION Yes

Schizophragma hydrangeoides var. *hydrangeoides* 'Roseum' is scented.

Chinese hydrangea vine is a useful climber for a shady, north-facing wall.

HOW THEY GROW

Rather than being categorized as climbers, hydrangea vines are technically deciduous shrubs that climb via adhesive aerial roots (*see p.35*), with which they cling to surfaces. They can be grown over pergolas and arches and up walls and even trees. They differ from climbing hydrangea because the frill of larger petals around the edge of their flowerhead each includes only one bract (modified leaf) rather than the four of true hydrangea.

Big, cream to white, slightly scented flowerheads are borne on Japanese hydrangea vine (*S. hydrangeoides*), which can grow to 8m (26ft) high and 1.5m (5ft) wide. The dark green foliage turns butter-yellow in autumn. Its cultivar *S.h.* var. *hydrangeoides* 'Roseum' is worth seeking out for its pink flowers, while *S.h.* var. *concolor* 'Moonlight' has attractive, silver-washed new foliage that changes to reddy orange in autumn.

Chinese hydrangea vine (*S. integrifolium*) has a similar height to Japanese hydrangea vine but is wider – it can spread up to 4m (13ft). Its leaves are longer and the flowers are bigger, with the bracts appearing in the flowerhead as opposed to just around the outside.

Hydrangea vines are shade-tolerant, but flower best in full sun, and thrive where their roots are in shade and their stems are in sun. Although fully hardy down to −10°C to −15°C (14–5°F), they benefit from a sheltered position. They grow in any moderately fertile, moist but well-drained soil – though Japanese hydrangea vine dislikes chalky soils. Plant a minimum of 60cm (24in) from the wall, tree, or other support, and tie in the stems until the plant is established and self-clinging via its adhesive aerial roots.

HOW TO PRUNE

Hydrangea vines do not need any regular pruning other than to remove dead, damaged or diseased growth. Once established, they can be trimmed when necessary to keep them to their allotted space, in spring. Prune out any damaged growth and reduce the stems and sideshoots to the required size.

RENOVATION PRUNING Renovate an overgrown hydrangea vine in stages over two or three years, rather than all at once. Every spring, cut back one-third of the oldest stems to the base, and shorten the others by one-half.

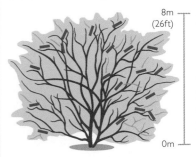

8m (26ft)

0m

In spring, shorten sideshoots and stems to fit the allocated space.

SPIRAEA *SPIRAEA*

These low-maintenance, mostly deciduous shrubs vary in size and growth habit, but all are appreciated for their profuse clusters of pretty flowers in spring or summer, in shades of white, pink, or red. They are unfussy and bring a flush of colour to a border or rock garden.

PLANT TYPE Deciduous and semi-evergreen shrubs
HEIGHT Up to 4m (13ft)
SPREAD Up to 2.5m (8ft)
FLOWERS ON Last year's and current year's stems, in spring or summer
LEAF ARRANGEMENT Alternate
WHEN TO PRUNE Early flowerers and hedges after flowering; late flowerers in early spring
RENOVATION Yes

HOW THEY GROW

Spiraeas come in many different forms, from upright shrubs with arching growth to dense and rounded, or compact, bushes. Many are spreading, suckering, or clump-forming, and vigorous, growing to 4m (13ft) high, while some reach only 50cm (20in). They are usually grown as specimen shrubs in a border, low-maintenance shrubbery, or woodland edge planting, but can also be used for informal hedging or ground cover. Spiraeas are of benefit to wildlife, including pollinators like butterflies.

Their greatest asset is their tiny, cup- or bowl-shaped flowers, with five petals, bunched together in clusters. Some species bloom in spring, while others do so in summer. Early-flowering types such as S. *prunifolia* typically bear flowers on cascading woody stems, while later-blooming species such as S. *canescens* are more likely to produce flowers on the current season's growth, at the branch tips. Spring-flowering bridal wreath (S. 'Arguta') reaches 2.5m (8ft) high and has gorgeous arching stems smothered in a foam of white flowers, while S. *japonica* 'Nana' is a dwarf type with pink flowers in summer, which creates a small mound to 45cm (18in). Unusually, S.j. GOLDEN PRINCESS is mostly grown for its leaves, which open bronze, before turning yellow in summer, and red in autumn; it also carries pink flowers in late summer.

Spiraeas are very easy to grow, and require little care after planting. They prefer fertile, moist but well-drained soil, and grow in full sun or partial shade, but flower best in a sunny spot. Although fully hardy, some benefit from a more sheltered position.

HOW TO PRUNE

EARLY-FLOWERING TYPES Prune spring bloomers that flower on the previous season's wood after flowering. Remove

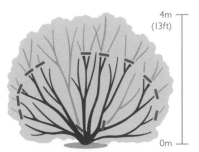

On early-flowering types, trim back growth and cut out some old stems.

dead, diseased, and damaged wood, and shorten flowered shoots back to strong buds lower down. Cut out up to one-quarter of the oldest stems to encourage fresh growth from the base.

LATE-FLOWERING TYPES Prune summer-blooming spiraeas that flower on the current year's growth in early spring. Cut back all stems each year to a low permanent framework.

HEDGES Trim to the required size immediately after flowering. Try to keep the natural loose shape rather than attempting a crisp formal finish.

RENOVATION PRUNING Spiraea responds to hard pruning to rejuvenate it. Cut all stems right back to about 15cm (6in) from the ground in winter or early spring. Thin out the resulting vigorous fresh growth, selecting the best shoots to grow on. The shrub may not flower for several seasons after this drastic renovation treatment.

Summer flowers are borne on low spreading *Spiraea japonica* 'Nana'.

LILAC *SYRINGA*

Lilacs are valued mainly for their conical flowerheads packed with intensely fragrant blooms ranging from pink, violet-blue, and white to pale yellow, which appear as spring turns to summer. For the rest of the year, these deciduous trees and shrubs tend to fade into the background in a garden.

PLANT TYPE Deciduous small trees and shrubs
HEIGHT Up to 7m (23ft)
SPREAD Up to 4m (13ft)
FLOWERS ON Last year's stems, in late spring–early summer
LEAF ARRANGEMENT Opposite
WHEN TO PRUNE After flowering
RENOVATION Yes for common lilac

HOW THEY GROW

Lilacs can be bushy, vigorous, upright or spreading shrubs, or small, often multi-stem trees, and are typically grown as specimen plants or at the back of a border. They can be a good choice for slopes, as part of a shelterbelt, and for wildlife, and are also able to cope with urban conditions and new-build sites.

Their crowning glory is the spring show of highly scented, pyramid-shaped panicles of tiny tubular flowers. Common lilac (*S. vulgaris*) and its cultivars are the most suitable types for gardens. *Syringa vulgaris* 'Madame Lemoine' has double white flowers, while *S.v.* 'Sensation' bears white-rimmed, purple-red blooms and *S.v.* 'Primrose' produces creamy yellow buds that open white. All of these flower from late spring into early summer and can grow to around 4m (13ft) high and wide.

Korean lilac (*S. meyeri* 'Palibin') carries mauve-pink flowers and is compact at a much more diminutive 1.5m (5ft) high and wide, making it perfect for a small garden. Another good choice for a smaller space or a container is *S.* 'Red Pixie', which has red buds that open pink, and grows to just 1.8m (6ft).

Most lilacs flower only once a year. However, the Bloomerang Series (such as *S.* BLOOMERANG PINK PERFUME and *S.* BLOOMERANG DARK PURPLE) will repeat-flower through the season, while *S. pubescens* subsp. *microphylla* 'Superba' bears one big flush of its pink blooms in spring, then several lesser ones.

Lilacs thrive in full sun, in fertile, humus-rich, neutral to alkaline soil that is well-drained, but they will still grow in poor soil or clay. Although fully hardy, they perform best when planted in a sheltered position, as late frost can damage new growth.

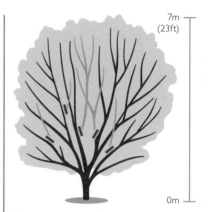

7m (23ft)

0m

Remove crossing shoots and older stems in the centre to enhance airflow.

HOW TO PRUNE

Apart from light pruning after flowering, to remove the spent heads, these trees and shrubs need only dead, diseased, damaged, and crossing growth removed. However, if you want to reduce the size of a lilac, you can shorten its flowered shoots back to a healthy bud lower down, and thin out up to one-quarter of the oldest stems, to enable air circulation through the centre of the plant.

RENOVATION PRUNING Common lilac and its cultivars are tolerant of being cut back hard after flowering. Either chop down all the stems at one time to 30cm (12in) from the base, or else stage this renovation over two or three years, removing up to one-third of the oldest stems each year.

Syringa **'Red Pixie'** is a small shrub with highly fragrant blooms.

Large, scented, double flowers adorn *Syringa vulgaris* 'Madame Lemoine'.

TAMARISK *TAMARIX*

A prime candidate for windy seaside gardens is tamarisk, which is tough as well as drought-tolerant. It is beloved for its wispy foliage and clouds of tiny, pink or white blooms covering its long and slender stems. These show-stopping flowers are attractive to wildlife and are sometimes scented.

PLANT TYPE Deciduous trees and shrubs
HEIGHT Up to 5m (16ft)
SPREAD Up to 5m (16ft)
FLOWERS ON Last year's and current year's stems, in late spring or summer–autumn
LEAF ARRANGEMENT Alternate
WHEN TO PRUNE Early flowerers after flowering; late flowerers in spring
RENOVATION Yes

Tamarix ramosissima **'Pink Cascade'** has fluffy pink plumes and feathery foliage.

HOW THEY GROW

Tamarisk is a spreading shrub or small tree with an open habit and arching bushy growth. The grey-green foliage is needle-like, and changes to orange in autumn, while the feathery flower plumes are borne all down the length of the dark-coloured shoots.

There are two types of tamarisk: early flowering (in spring) and late flowering (in summer). Typical of the former type is *T. tetrandra*, which grows to 3m (10ft) or more, flowers on old wood, and produces its pale pink blooms on purple-black branches in late spring.

A late-flowering type is *T. ramosissima* and its cultivars, which can reach 5m (16ft) high and wide, and tend to flower in summer to autumn on new growth; *T.r.* 'Hulsdonk White' bears white blooms, while those of *T.r.* 'Rosea' and *T.r.* 'Pink Cascade' are pink; *T.r.* 'Rubra' has dark pink flowers and red stems.

Tamarisk is exceptionally resilient, able to grow even in sandy, saline, dry soil and in windswept, salt-sprayed areas. As a result, it is often planted in coastal regions. This ability to survive harsh conditions means it is considered invasive in some regions. Grow it in acid to neutral, well-drained soil in full sun. In inland gardens, the soil should be moist but well-drained. Although fully hardy to at least −10°C (14°F), away from coastal areas, tamarisk requires shelter from cold drying winds.

HOW TO PRUNE

After planting, cut back young plants hard, almost to the ground, to encourage them to develop bushy growth. Once established, prune regularly to prevent plants from growing top-heavy and becoming damaged. Trimming each year helps keep them compact and improves flowering. To grow as a tree, lift the crown of established plants in winter by removing the lower lateral stems (see p.31).

EARLY-FLOWERING TYPES For those that flower in spring, prune immediately after flowering. Remove damaged and

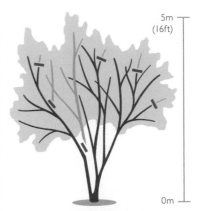

5m (16ft)

0m

Shorten flowered shoots and thin out some of oldest stems on spring bloomers.

crossing growth, and cut back the flowered stems to strong growth lower down. Thin out up to one-quarter of the oldest stems to encourage fresh growth from the base.

LATE-FLOWERING TYPES For those that flower in summer and into autumn, prune in spring, just before fresh growth begins. Cut down all stems to a permanent framework within 60cm (24in) of ground level, depending on how you wish it to grow and what size you require.

RENOVATION PRUNING Leggy overgrown tamarisk plants usually respond well to hard pruning. Cut all stems down to near ground level in spring. However, you may lose flowers for a season or two after this drastic renovation treatment.

YEW *TAXUS*

The tightly knit, linear, emerald-green leaves of yew make it perfect for trimming into dense, clean-edged hedges and other shapes, which create welcome structure in the garden. Yew is unfussy, too, coping well with a range of difficult growing conditions.

PLANT TYPE Evergreen trees and shrubs
HEIGHT Up to 12m (39ft)
SPREAD Up to 8m (26ft)
FLOWERS ON Last year's stems, in late winter–early spring
LEAF ARRANGEMENT Spiral or two-ranked
WHEN TO PRUNE Late summer–early autumn
RENOVATION Yes

HOW THEY GROW

This evergreen conifer is used as formal hedging and for clipping into topiary shapes. Common yew (*T. baccata*) is a slow-growing, bushy shrub or tree with a naturally conical to rounded form when cultivated as a free-growing plant. It has rows of narrow, needle-like leaves, which are deep green on top and grey-hued underneath. Yew tolerates exposed sites, deep shade, dry soil, and urban conditions including air pollution. When left unpruned, it will eventually grow into a tree with textured, red-brown bark. There are also low-growing types, such as mat-forming *T.b.* 'Repens Aurea', which is useful as ground cover, as well as upright columnar forms like golden Irish yew (*T.b.* 'Fastigiata Aureomarginata').

Although tough and hardy, the fastigiate types prefer a sheltered spot. Yew grows in any well-drained soil, in full sun, partial shade, or full shade. It dislikes boggy waterlogged conditions.

Almost all parts of yew are very toxic if ingested, so wear gloves when handling it. Tidy up clippings carefully, because the leaves are also dangerous to eat for pets and livestock. Warn people, especially children, to stay away from the fruit, as the seed inside the red casing is deadly poisonous.

HOW TO PRUNE

FREE-GROWING TYPES These require only the cutting out of dead, damaged, and diseased growth, and any overlong or wayward shoots, in late summer or early autumn.

HEDGES AND OTHER CLIPPED SHAPES
Hedging, topiary, and yew grown as a specimen upright shrub need annual attention to stay neat. On hedges and upright shrubs, use hedge trimmers to clip back to the desired size, and trim topiary with topiary shears to keep it in shape. Allow upright shrubs to gain the height you want, and a little extra, before beginning to trim the top. On the sides, clip new shoots back to maintain the upright growth and tall columnar form.

While plants are getting established, you can encourage bushier denser growth by pruning twice a year, in spring and again in late summer or early autumn. Once established and to the size and shape desired, routine pruning once a year in late summer or early autumn is enough to keep a crisp finish.

RENOVATION PRUNING Yew can grow leggy and bare at the base when neglected, but will resprout from old wood and can be cut back very hard during spring. Some gardeners prune overgrown or gappy specimens down to 15cm (6in) from the ground, but yew's slow growth means it can take years to recover. It is better to take out no more than one-third of the top-growth at one time.

Similarly, hedges should be restored over several years, cutting the top back in year one, then reducing the width of one side the following spring, and the other one the spring after that.

Common yew is excellent for creating complex topiary shapes.

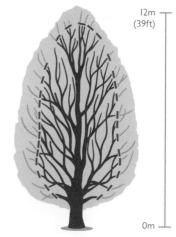

12m (39ft)

0m

Trim back new growth on columnar shrubs to retain the shape.

VIBURNUM *VIBURNUM*

Not only are viburnums easy to grow, but these shrubs also have many wonderful attributes that make them year-round superstars of the garden. They produce often scented clusters of pink or white flowers, followed by blue, red, or black berries, as well as wonderful shades of autumn foliage.

PLANT TYPE Evergreen, semi-evergreen, and deciduous shrubs
HEIGHT Up to 4m (13ft)
SPREAD Up to 4m (13ft)
FLOWERS ON Last year's and current year's stems, in autumn–spring, or spring
LEAF ARRANGEMENT Opposite
WHEN TO PRUNE Evergreens in late winter –early spring; deciduous after flowering
RENOVATION Yes

Viburnum plicatum f. tomentosum 'Mariesii' has a white-petalled outer ring.

Guelder rose (*V. opulus*) develops colourful berries and leaves in autumn.

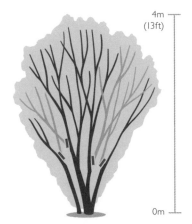

4m (13ft)

0m

Prune out damaged growth and any crossing or rubbing shoots and stems.

HOW THEY GROW

Viburnum is a diverse genus, its species having very different characteristics. Sizes range from low-growing types to large shrubs up to 4m (13ft) tall and 6m (20ft) spread, but most grow to around 2.5m (8ft). The growth habit also varies from spreading, open, and upright, to compact and rounded or doming. Viburnums are a good choice for shrub plantings, as specimen shrubs, or in a woodland garden – usefully, these low-maintenance plants are also deer-resistant.

Another point of interest is their abundant blooms, sometimes fragrant, which appear in white, cream, and pink, often with contrasting-colour buds. The foliage is in different shapes and sizes, but the best autumn colour is found on the deciduous types. Many species also develop berries after flowering, which are a boon for wildlife like birds.

Useful garden types include *V. × bodnantense* 'Dawn', a vigorous deciduous shrub with scented pink flowers opening from red buds, from autumn to spring, and deciduous *V. plicatum* f. *tomentosum* 'Mariesii', notable for its distinctive tiers of horizontal branches and lacecap-like, white flowers in late spring. Evergreen laurustinus (*V. tinus*) produces its white flowers from late winter to spring, followed by dark blue berries.

Viburnums are hardy and adaptable shrubs that grow in any moderately fertile, moist but well-drained soil, in full sun or partial shade. However, evergreens benefit from a sheltered site out of cold drying winter winds.

HOW TO PRUNE

DECIDUOUS TYPES Prune non-berrying plants immediately after flowering, cutting overlong stems back to fit the allotted space and shape, if required. Thin out congested plants by removing up to one-third of the oldest stems.

EVERGREEN TYPES Minimal pruning is required other than the routine removal of dead, damaged, or diseased wood and any crossing or rubbing shoots in late winter or early spring. If necessary, take out overlong or wayward stems that spoil the symmetry of the plant.

RENOVATION PRUNING Deciduous types and evergreen laurustinus and its cultivars tolerate hard pruning to rejuvenate them if they become overgrown or leggy. Cut back all stems to within 15–30cm (6–12in) of the base in late winter or early spring, before fresh growth starts.

ORNAMENTAL VINE *VITIS*

Ornamental vines – unlike grape vines – are cultivated mainly for their large, lobed or heart-shaped leaves, which often turn brilliant fiery shades in autumn, lighting up the garden. Some types have fruit and, although usually edible, it is small and best left on the plant for decorative effect.

PLANT TYPE Deciduous climbers
HEIGHT Up to 12m (39ft)
SPREAD Up to 4m (13ft)
FLOWERS ON Current year's stems, in summer
LEAF ARRANGEMENT Alternate
WHEN TO PRUNE Winter
RENOVATION Yes

HOW THEY GROW

These deciduous climbing shrubs self-cling with their leaf tendrils (see *p.35*). Being large and vigorous, they quickly clothe a wall or fence, when given support from a trellis or system of wires (see *p.36*), and are sometimes also grown up into trees.

Many ornamental vines develop vibrant autumn foliage colours, and it is this, rather than their unremarkable green flowers in summer, that is the primary reason for growing them. The blooms, however, are attractive to bees, and are followed in some cases by bunches of dark spherical fruit.

Vitis 'Brant' has three-lobed, green leaves that change to purple and ruby-red with green veining in autumn.

Small green grapes appear in summer and turn black as the season changes. Although edible, they are very full of seeds, so should probably be left for decoration – and the birds.

Crimson glory vine (*V. coignetiae*) is fast-growing with heart-shaped, green leaves with felty brown undersides. The foliage turns bright crimson in autumn. The grapes sometimes produced are not tasty. Another popular cultivar is *V. vinifera* 'Purpurea', with its leaves that darken to purple over the season.

Ornamental vines prefer a position in full sun, where they will produce the best autumn colour, but they also grow in partial shade. They are fully hardy to around –15°C (5°F), but should be planted in a sheltered spot to avoid frost damage to fresh growth. Grow

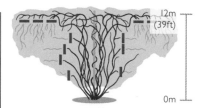

Trim excessive growth back to buds farther back on shoots and stems.

in fertile, well-drained soil that is neutral to alkaline. Wear gloves when handling vines, to avoid skin irritation.

HOW TO PRUNE

After planting, guide the stems to the support and tie in as they grow (see *pp.38–39*). Vines do not require regular pruning, although you may need to restrict the size of these vigorous growers to fit their allotted space. Cut overlong shoots back to a healthy leaf node lower down the stem, in winter. To maintain a neater, more formal look, you can also reduce lateral stems to within three or four buds of the main stems.

RENOVATION PRUNING If your ornamental vine has become overgrown and needs to be renovated, you can cut it back hard, to about 30cm (12in) from the ground during winter. It will respond with fresh leafy growth but may not flower or fruit for a few years after this. Tie the new stems to the support, to create a basic framework, and in the second year thin these out to promote the best stems.

Vitis vinifera **'Purpurea'** leaves turn from green to bronze-pink to dark purple.

Vigorous crimson glory vine will quickly grow up into a tree.

WEIGELA *WEIGELA*

These low-maintenance, easy-to-grow shrubs carry clusters of showy, bell-shaped or tubular flowers, sometimes scented, in late spring or early summer in colours from red and pink to white and yellow. Some types also display eye-catching foliage in shades of bronze, purple, or yellow, or variegated.

PLANT TYPE Deciduous shrubs
HEIGHT Up to 4m (13ft)
SPREAD Up to 4m (13ft)
FLOWERS ON Last year's stems, in late spring–early summer
LEAF ARRANGEMENT Opposite
WHEN TO PRUNE After flowering
RENOVATION Yes

HOW THEY GROW

Weigela are bushy deciduous shrubs that range in size from dwarf types like *W.* Pink Poppet at 40cm (16in) to large specimens such as *W. coraeensis* 'Alba', which can reach up to 4m (13ft) in height and spread. Most weigela, however, grow to around 2.5m (8ft). The growth habit varies from upright or arching to mounding or spreading. Weigela is a good choice for adding interest to a mixed border or shrub planting, and also works well in the open aspect of a woodland garden. Smaller cultivars are useful when planted in a row to edge a path or bed.

These shrubs develop an excellent selection of leaf colour, from variegated green and cream to golden and bronze foliage, and purple to near-black, while their abundant clusters of small colourful flowers are particularly attractive to butterflies. *Weigela* 'Snowflake' has plain green leaves and prolific, pure white flowers, while *W. florida* 'Alexandra' produces deep bronze to purple foliage that darkens over time, and dark pink flowers. *Weigela praecox* 'Variegata' has cream-edged, green leaves and fragrant pink flowers with yellow splotches.

Weigela are unfussy plants, tolerating a range of conditions, and are very easy to look after, as well as being deer-resistant. All are fully hardy, but most prefer a sheltered position, though a few (such as *W.* 'Mont-Blanc') can cope with exposed sites. They thrive in full sun, but still grow in partial shade. Those with non-green foliage will colour up better in sun, while variegated types may perform better with some afternoon shade. Grow in any slightly acidic to slightly alkaline soil that is moist and well-drained.

HOW TO PRUNE

Weigela flowers on the previous year's growth, so should be pruned after blooming. Shorten flowered stems back to strong buds or shoots lower down the stem. Every few years, to encourage fresh growth from the base on large established specimens, cut back up to one-third of the oldest stems to the ground. Dwarf types do not require much, if any, pruning, apart from deadheading, and the removal of dead or damaged growth, if necessary. Variegated types may develop plain green shoots; remove these at once, to prevent the plant "reverting" (see p.10).

RENOVATION PRUNING Weigela can become lopsided, bare-legged, and unshapely if not pruned regularly, or if grown in full shade. Fortunately, it responds well to severe restorative pruning to reduce or rejuvenate an old, overgrown, or neglected plant. Shorten all stems to around 30cm (12in) from the base in late winter or early spring, when the plant is dormant. Leafy growth will resprout but you will not get flowers the following season.

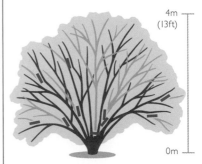

4m (13ft)

0m

Prune out damaged growth as well as some of the oldest stems.

The bright golden foliage on *Weigela* 'Looymansii Aurea' enhances its flowers.

WISTERIA *WISTERIA*

Wisteria is a large vigorous climber with an abundance of feather-like leaves. It is beloved for its long, pendent, grape-like clusters of blue, purple, pink, or white flowers – similar to those of peas, and sometimes fragrant – hanging down gracefully in late spring or early summer.

PLANT TYPE Deciduous climbers
HEIGHT Up to 12m (39ft)
SPREAD Up to 8m (26ft)
FLOWERS ON Last year's stems, in late spring–early summer
LEAF ARRANGEMENT Alternate
WHEN TO PRUNE Midsummer and mid-late winter
RENOVATION Yes

HOW THEY GROW

These deciduous climbers, with their twining stems, are often grown on trellis or wires against a wall or strong fence, or over structures like an arch or pergola. They can also be trained into a tree or as a standard. When left to grow unchecked, they become dense, heavy, and very large, often growing into gutters and roof spaces. Therefore, before planting, you should consider their positioning carefully and provide a sturdy support if possible. Regular pruning also helps to manage the weight and spread. It is well worth braving this strong-growing tendency, however, so you can enjoy the springtime display of cascading flowers along the shoots, followed by long, velvety, bean-like seed pods.

Wisteria is hardy but grows best in full sun in a south- or west-facing position sheltered from cold drying winter winds and late frosts, which can damage the flower buds. Plant in any fertile, moist but well-drained soil.

Be aware that all parts of the plant are toxic if ingested.

HOW TO TRAIN

AGAINST A WALL OR FENCE Wisteria can be allowed to clothe surfaces for a loose romantic look simply by tying the stems to a support such as trellis and allowing it to romp at will. Chinese wisteria (*W. sinensis*) has short blooms, about 30cm (12in) long, which mean it

is ideal for growing against a wall. To grow wisteria on a wall or very strong fence in a formal manner, it is best to train it in the espalier style using horizontal wires (see *pp.38–39*). After planting, cut the main stem down to about 90cm (36in), then select lateral stems as they develop, and secure them to the horizontal wires. In winter, shorten the sideshoots that grow off the lateral stems to two or three buds, and remove other laterals. Repeat every year until the plant is established, to create a basic framework.

ON A FREESTANDING STRUCTURE Wrap wires around the pergola or arch posts (see *p.37*). Then train and tie the developing stems to the wire and across the top of the structure. *Wisteria*

floribunda 'Multijuga' has flowers 1.2m (4ft) long, making it excellent for growing on such a structure.

OVER A TREE Grow wisteria only into a large and strong tree because it will damage a small tree. Be aware, however, that training and pruning it once established will probably not be possible. Plant at least 1m (3ft) from the tree trunk in full sun. Guide the wisteria stems towards the trunk, tying them on gently for the first few seasons until established and far enough into the tree to support themselves.

AS A STANDARD Let a single stem grow until it reaches the top of the stake supporting it. Then cut its top off in order to encourage lateral stems to

In summer, wisterias send out long sprawling shoots in search of supports.

Shortening long summer shoots puts the plant's energies into next year's flowers.

Winter pruning year on year gradually creates stubby clusters known as "spurs".

develop. Tip-prune the new growth to encourage the head of the standard to fill out and become bushy.

HOW TO PRUNE

Prune wisteria twice a year, during midsummer and in mid- or late winter, to control its growth and encourage more flowers. The key to getting the most from your plant is to understand how it grows and flowers. Think of the plant in three parts: the main stem; the lateral stems, which grow off the main stem; and the sideshoots, which grow from those lateral stems. Because the blooms appear on the sideshoots,

A magnificent wisteria in its prime: the seeds these flowers form are toxic to birds, so summer pruning after flowering won't deprive them of food.

you want to encourage as many sideshoots as possible to develop all along the lateral stems. Therefore, get rid of a lot of the leafy growth.

WALLS AND FREESTANDING STRUCTURES Begin maintenance pruning only when the wisteria has covered the structure it is growing on. In midsummer, after the plant has flowered, shorten the long, whippy, young, leafy sideshoots back to within five to six leaves, or around 30cm

(12in). This not only promotes air circulation and allows light on to the wood, but also encourages flowering spurs to develop. Then in mid- or late winter, shorten these same growths back to two to three buds, or about 10cm (4in). At the same time, cut back lateral stems to within bounds.

STANDARDS In midsummer, once the basic standard shape has been established, cut back to seven leaves any sideshoots that are not needed to shape the head. In mid- or late winter, shorten them again to within 2.5cm (1in) of the lateral stems.

RENOVATION PRUNING Old overgrown wisteria, or a plant with big woody branches that has become top-heavy or has damaged its support, can be pruned hard. In mid- or late winter, cut branches and smaller stems right back to the size you require – even back close to the base, as long as you don't cut below the graft point on grafted types. This drastic renovation will stimulate lots of new growth that will have to be thinned and trained into the plant's support over several seasons, like a newly planted wisteria (see *opposite*). There will be fewer or no flowers for a year or two.

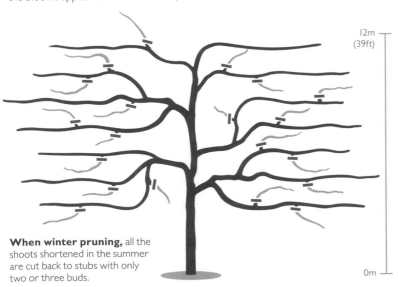

12m (39ft)

0m

When winter pruning, all the shoots shortened in the summer are cut back to stubs with only two or three buds.

INDEX

Bold text indicates a main entry
for the subject.

Author Stephanie Mahon

PUBLISHER ACKNOWLEDGMENTS

DK would like to thank Oreolu Grillo and Sophie State for early spread development for the series, and Margaret McCormack for indexing. Thanks to Zia Allaway, Jo Whittingham, and Ian Spence for additional text and research.

PUBLISHER ACKNOWLEDGMENTS

DK would like to thank Oreolu Grillo and Sophie State for early spread development for the series, and Margaret McCormack for indexing.

PICTURE CREDITS

The publisher would like to thank the following for their kind permission to reproduce their photographs:

Alamy Stock Photo: Martin Hughes-Jones 6c; agefotostock 19tc; Nigel Cattlin 23br; Wiert Nieuman 25cr; Avalon/Photoshot License 30tr; Kay Roxby 33bc; Tim Gainey 34tr; RF Company 34br; Tim Gainey 35tl; Elizabeth Whiting & Associates 36bl; Tim Gainey 38c; Avalon/Photoshot License 39tl; Alexey Solodov 42tr; BIOSPHOTO 45tc; blickwinkel 54cb; shapencolour 66c; Dave Bevan 70br; Jolanta Dąbrowska 71cb; Nigel Cattlin 74bl; Michael Russell 77tl; garfotos 80bl; John Glover 82tc; Zoonar GmbH 93tl; BIOSPHOTO 124bl; Avalon/Photoshot License 132cl; Müller / McPhoto 141tr.

Dorling Kindersley: 123RF.com / Leonid Ikan 11tl; 123RF.com / Tawatchai Prakobkit / prakobkit 11c; Alan Buckingham 20cl, 20c, 20cr; Brian North / Waterperry Gardens 8br, 27tr, 49br; Dreamstime.com / John Pavel 56bl; Wiertn 80tr; Drew Lehman / Dlehman97 87tl; Mark Winwood 26tr, 84bc, 99tc; Mark Winwood / Capel Manor College / Irma Ansell 24bc; Mark Winwood / Crug Farm 51cr, 91bc, 92bl; Mark Winwood / Downderry Nursery 100bl; Mark Winwood / Lullingstone Castle, Kent 115cl, Mark Winwood / RHS 9tc, 77br; Mark Winwood / RHS Wisley 8bc, 9bl, 10bl, 13bc, 13br, 15bc, 43bl, 44tl, 45tr, 46cl, 52bl, 55bl, 57tl, 57c, 70bl, 72bl, 73tl, 74cb, 75cl, 75c, 75br, 79tl, 79cb, 85bl, 90bl, 90bc, 91tl, 94tl, 101cl, 106bl, 110cl, 114bl, 116cl, 117bl, 118bl, 119tc, 122bl, 122cr, 124tr, 125br, 128tl, 134bl, 134bc; Peter Anderson 16cr, 17cl, 25cl, 26bl, 30bl, 31tc, 50c, 53cl, 53cb, 123tl, 136bl; RHS Wisley 120c; Savill Garden, Windsor 114bc; Steve Hamilton / Cambridge Botanic Gardens 29tr.

GAP Photos: Jonathan Buckley 18tr; Jacqui Dracup 33cl; Paul Debois 33br.

Illustrations by Cobalt id.

Cover images: Front: Dreamstime.com: Dmitri Maruta

All other images © Dorling Kindersley

Produced for DK by COBALT ID
www.cobaltid.co.uk

Managing Editor Marek Walisiewicz
Editor Joanna Chisholm
Managing Art Editor Paul Reid
Art Editor Paul Tilby

DK LONDON
Project Editor Amy Slack
Managing Editor Ruth O'Rourke
Managing Art Editor Christine Keilty
Production Editor David Almond
Production Controller Stephanie McConnell
Jacket Designer Nicola Powling
Jacket Co-ordinator Lucy Philpott
Art Director Maxine Pedliham
Publishers Mary-Clare Jerram, Katie Cowan

First published in Great Britain in 2021 by Dorling Kindersley Limited DK, One Embassy Gardens, 8 Viaduct Gardens, London, SW11 7BW

A CIP catalogue record for this book is available from the British Library.
ISBN: 978-0-2414-5860-0

Printed and bound in China

For the curious
www.dk.com

Pruning & Training